The Man Who Saw Through Time

ALSO BY LOREN EISELEY

THE IMMENSE JOURNEY
1957
DARWIN'S CENTURY
1958
THE FIRMAMENT OF TIME
1960
THE UNEXPECTED UNIVERSE
1969
THE INVISIBLE PYRAMID
1970
THE NIGHT COUNTRY
1971
NOTES OF AN ALCHEMIST
1972

The Man Who Saw Through Time

Loren Eiseley

REVISED AND ENLARGED EDITION OF
Francis Bacon and the Modern Dilemma

CHARLES SCRIBNER'S SONS
NEW YORK

The lines from Archibald McLeish, "Epistle to Be Left in the Earth,"
from *Collected Poems 1917-1952*, quoted on pages 78-79,
are used by permission of the publisher,
Houghton Mifflin Company, Boston, Mass.

Charles Scribner's Sons
Macmillan Publishing Company
866 Third Avenue, New York, NY 10022

Library of Congress Catalog Card Number 72-12150

ISBN 0-684-13285-0

Macmillan books are available at special discounts for bulk purchases
for sales promotions, premiums, fund-raising, or educational use.
For details, contact:

Special Sales Director
Macmillan Publishing Company
866 Third Avenue
New York, NY 10022

15   14   13   12   11

Printed in the United States of America

To
Francis Bacon
and Sir Thomas Meautys
his faithful secretary
who erected his monument
and chose in death
to lie at his feet
sharing honor and disgrace
this tribute
from one who
more than three centuries beyond their grave
is still seeking
the lost continent of their dream

For the history that I require and design, special care is to be taken that it be of wide range and made to the measure of the universe. For the world is not to be narrowed till it will go into the understanding (which has been done hitherto), but the understanding is to be expanded and opened till it can take in the image of the world.

—Francis Bacon, *The Parasceve*

Contents

Preface

On the occasion in 1961 of the four-hundredth anniversary of Francis Bacon's birth I had occasion to engage in several addresses at institutions of learning in the United States upon Bacon's life and contributions to science. I rapidly discovered that I was unwittingly assuming the role of attorney for the defense against sometimes extremely self-righteous public prosecutors who had been unduly influenced by Thomas Babington Macaulay's intemperate and acerbic treatment of Bacon in the nineteenth century, writings upon which many later treatments of Bacon were modeled.

I found myself embroiled, in fact, in sufficient controversy to make me wonder whether it was I who was threatened with the Tower and whether Parliament was in full cry upon my own derelictions. One critic even berated me for mentioning that Bacon had expressed himself upon the value of mathematics to science; others chose to criticize me, when certain of these lectures were published, for failing to write a full-length biography—something that was not the purport of my efforts in that anniversary year; my intention was rather to do honor to a great scholar and seer.

As I come to reprint these papers I am touched to discover that much which I defended has been, in the interim, accepted. Hugh Trevor-Roper has commented upon the Puritan vulgarization of Bacon's thought, which was to be followed by the Victorian vulgarization of Macaulay. As the historian Margery Purver has remarked, "the Victorian image of Francis Bacon, distorted and inconsequential, continues to dominate historians." I will not further explore what Miss Purver has reviewed with consummate grace. Rather, as an anthropologist, I would like to emphasize that for reasons involving the "decay of nature" the average Elizabethan vocabulary was, like that of early man, inadequate to treat of events remote in future time. Bacon, by contrast, was attempting to project for the masses a new definition of culture and inventiveness extending into the remote future. Semantically it involved as difficult a task as Darwin was later to encounter in his attempts to explain natural selection.

Again, it must be realized that Bacon, more fully than any man of his time, entertained the idea of the universe as a problem to be solved, examined, meditated upon, rather than as an eternally fixed stage upon which man walked. Instead, there was a lurking novelty within it which forecasts the process philosophers of the late nineteenth and the twentieth centuries. He counsels us, indeed, that we must distinguish between the normal *course* of nature, the *wanderings* of nature, which today we might associate with the emergence of the organically novel, and, finally, the "art" that man increasingly exerts upon nature and that results, in turn, in the innovations of his cultural world, another kind of hidden potential in the universe. All of these categories he subsumed under "natural history."

For those who would see in Bacon only the worst aspects of modern industrialism, I should like to close with his admonition about man. "Let us now come to that knowledge . . . which is the knowledge of ourselves, which deserves the more accurate handling in proportion as it touches us more nearly. This knowledge is for man the end and term of knowledge." Bacon, in other words, was again too farsighted to remain enclosed in the circle of mechanical technology. He was truly a man for the ages and his insight soars beyond us still.

—Loren Eiseley

The Man Who Saw Through Time

1

An Age of Violence

In January of 1561 a son was born to Nicholas Bacon, Lord Keeper to Elizabeth I. In twelve years this bright, grave child, Francis, would be called by Elizabeth her little Lord Keeper, but all his life she would deny him great office, as one denies, yet counsels with a wizard, and all his life poverty and ill fortune would dog him in the midst of luxury. Yet it is this man who first fully visualized in all its splendor the "invention of inventions"— the experimental method which would unlock the riches of the modern world.

The most curious aspect of the technological environment which surrounds us today is one we rarely think about—namely, that it exists. How did it arise and why? We define ourselves vaguely as *Homo sapiens,* the wise, and we assume, if we think about our surroundings at all, that man's innate wisdom has, in the course of time, automatically produced the scientific world we know. Yet the archaeologist would be forced to tell us that several great civilizations have arisen and vanished without the benefit of a scientific philosophy. Similarly, Western society, down to the last three centuries or so, betrays but feeble traces of that type of thinking known today as "scientific," with its emphasis upon experiment and dispassionate observation of the natural world.

There is only one great exception in Western thought— Greek philosophy and Greek science before that tiny but enlightened world was destroyed by internecine conflict and the expanding power of the Roman Empire. In other words, this wise creature, man, has rarely shown any penchant for science and would much rather be left to his uninhibited dreams and fantasies.

Scientific thought demands some kind of unique soil in which to flourish. It has about it the rarity of a fungus springing up in a forest glade only to perish before nightfall. Perhaps, indeed, its own dynamism contains its doom. Perhaps the tendency of science to fragment and crumble also partakes of the qualities of the mushroom. This much at least we know: science among us is an *invented* cultural institution, an institution not present in all societies, and not one that may be counted upon to arise from human instinct.

Science is as capable of decay and death as any other human activity, such as a religion or a system of government. It cannot be equated with individual thought or the unique observations of genius, even though it partakes of these things. As a way of life it has rules which have to be learned, and practices and techniques which have to be transmitted from generation to generation by the formal process of education. Neither is it technology, although technology may contribute to science, or science to technology. Many lost civilizations—Roman, Mayan, Egyptian—had great builders, whether of roads, aqueducts, temples, or pyramids. Their remains show enormous experience of transmitted and improved techniques, but still these are not precisely within the true domain of science.

Science exists only within a tradition of constant experimental investigation of the natural world. It demands that every hypothesis we formulate be subject to proof, whether in nature or in the laboratory, before we can accept its validity. Men, even scientists, find this type of thinking extremely difficult to sustain. In this sense science is not natural to man at all. It has to be learned, consciously practiced, stripped out of the sea of emotions, prejudices, and wishes in which our daily lives are steeped. No man can long endure such rarefied heights without descending to common earth. Even the professional scientist frequently confines such activity to a specific discipline and outside of it indulges his illogical prejudices.

To introduce the concept of the scientific method into the world at large as a way of life, therefore, is a more arduous and difficult task than merely to conceive of its philosophical possibility. Even today, when scientific achievements surround us on every hand and the textbooks of our schools bulge with illustrative experiments, men are, in the mass, still emotional and resistant to fact, particularly in their political and social thinking. We are always more willing to accept mechanical changes in an automobile than to revise, or even to examine our racial prejudices, to use just one painful example. We are more willing to swallow a pill that we hope will relax our tensions than to make the sustained, conscious effort necessary to alter our daily living habits.

The interest and significance of Francis Bacon, the little boy born at Elizabeth's court, lies in this: he played a powerful role in getting English society to swallow, figuratively, a pill—the pill of science. It took most of his adult life and the patient endurance of the court's and the world's contempt, before, a few generations later, society finally gulped down its medicine and turned to look in the mirror.

Since the dawn of the scientific world is a strangely unique, almost unnatural one, the life and times of the great statesman who played a major part in the half-light of that spectral morning will be of perennial interest so long as science and its world endure. Of all those who dreamed in secret, experimented but confined their endeavors, Francis Bacon alone walked to the doorway of the future, flung it wide, and said to his trembling and laggard audience, "Look. There is tomorrow. Take it with charity lest it destroy you."

Time dulls the horrors of the past. A traditional monarchy like that of Britain has much in its history which both subjects and later monarchs desire to forget. We remember Elizabeth as an adept ruler of men; we remember her sea rovers and the destruction of the Spanish Armada. As the centuries recede the screams of tortured men sound far away; the fall of the headsman's axe is no longer heard upon the block. But to understand the Elizabethan world in which Francis Bacon rose to a Lord Chancellorship we must know the realities of that world.

Its bloody legacy came straight from Henry VIII. It was a world in which the English language was still being

shaped into the vehicle of great literature. It was also a
world of terror, corrupted by absolute power, a world
where the throne defined treason and where to rise was
also to invite one's fall, a world where rulers expected to
be struck in the dark and where the logical paranoid act
was to strike first. "Every public man in the England of
the Tudors and the Stuarts," writes one historian, "en-
tered on his career with the familiar expectation of possibly closing it on the scaffold."

Spies and informers swarmed everywhere. *Agents
provocateurs* promoted treason for pay. Bestial crowds
swarmed to torturings and hangings in public. The heads
of executed men withered on the city gates as object les-
sons. Withal, this was Shakespeare's world and Francis
Bacon's. It was the latter who was to write of himself and
his generation as wearing out days few and evil. It was he
who was to say, in weariness, that his soul had been a
stranger to his pilgrimage. It was he, the last great Elizabethan, who was to murmur when they came to summon
him back to the court of Charles I, "I have had enough of
that vanity."

All along the way he had done the will of princes, had
been a true servant of the state. Why was disgrace his re-
ward? The world forgives his treasonous, intemperate patron, Essex, who died under the axe. The world remem-
bers with favor the great freebooter, Raleigh, who ate his
heart out in the Tower and could not believe his age was
dead until he had tried the seas once more. These men
were truly Elizabethans. They died bloodily, as their age
demanded, and were understood.

Francis Bacon, by contrast, walks masked and cool through this age of violence. Traps snap on either side, associates perish, he remains. Even when he is caught between parliament and king, and a powerful enemy demands his imprisonment in the Tower—even then, he goes free, though robbed by a strange combination of events of his personal honor, his fortune, and his place at court.

Yet among all these plots and subplots a curious mythology lives on: that Francis Bacon "took a bribe"—a not very respectable thing for a judge to do, a thing to moralize upon in the safe seclusion of a modern study. Only recently the charge was raised again in conversation as a new legal book heaping invective upon him came up for discussion. Though Francis Bacon valued his good name and left it, when he knew the worst had come upon him, "to the next ages and to foreign nations," it is doubtful that he, or James I, or James's callous favorite, George Villiers, first Duke of Buckingham, could have foreseen how close that charge would stick after three hundred years.

The men of violence have been forgiven. A romantic halo envelops them. But the man who outlived the violence and who husbanded his power of survival in order to communicate a great secret, our age finds it oddly difficult to forgive. One wonders why. Perhaps it is because he was truly a stranger in his own age—a civilized man out of his time and place, dealing with barbarians and barely evading the rack and gallows in the process. It affronts our sense of dignity to see him bowing painfully to titled fools and rapacious upstarts, while presenting his books hopefully to learned men who scornfully fling them aside. He walks hesitantly toward us through history as though he could see our century but not reach it; he is out of place.

In a grim moment he whispers in *The Advancement of Learning* that one must consider how one's nature suits the state of the times. If one is out of place then one must walk "close and reserved." This is all we ever learn of the man except that he had one burning passion: to change the world through thought, through an "engine" he had devised. Otherwise there is nothing of himself; nothing, that is, save the cry of the painter Nicholas Hilliard, who wished he could have painted Bacon's mind, and the words of Ben Jonson, who spoke of his eloquence, and of another who remarked upon his generosity.

Bacon died, appropriately enough, in the midst of an experiment. He had gone out in a carriage on a winter day and decided suddenly to investigate the effects of cold in delaying putrefaction. He stopped his carriage, bought a hen from a cottage woman, and stuffed it full of snow. Immediately aware of having taken a chill himself, he sought the hospitality of a friend who lived nearby, and in that house he died.

It is symbolic that Bacon died in a borrowed bed. The century in which he found himself was equally borrowed, and he had no genuine place within it. He is Faust and something more. In some fashion he is ourselves, and we project upon Bacon the fear we have of what he has brought to us. His work is not a gift that can be recalled, and the more we come to fear the gift, the more hatred we extend to the giver of it. Before we can understand and therefore forgive the giver, we must understand the intention of his gift.

Rumor has it that Francis Bacon, his father's favorite child, the boy whose courtesy had caught the eye of the queen, played alone a great deal. Once he was found investigating the nature of echoes in some rocky spot, the story goes, and a likely enough incident it is, for what boy has not shouted into a well or bounced experimental sounds in reverberating places? But it is also likely that his intent, listening ear caught sounds not heard by ordinary, boisterous children. In the fey mind of this solitary, gifted youth it might have seemed as though fate whispered, or an echo sounded from some yet distant century. For Bacon, at least in his adult years, was to show an uncanny sensitivity to time.

In 1573 Francis, along with his brother Anthony, went to Cambridge. He was just twelve, unusually young even for that day of early university training. Cambridge at the time had sunk to a low intellectual level and Bacon did not linger long. The school was given over to bitter theological disputes. "Men of sharp wits," Bacon was later to describe his tutors, "shut up in the cells of a few authors, chiefly Aristotle, their Dictator." A strong admiration for the lost classic literature could not conceal the fact that at Cambridge learning was largely pretense, that all was of the past. Men endlessly wove and rewove a spider web of ideas derived from Greek and Roman sources.

At the age of fifteen the youth returned home, indifferent to his abandoned degree. In 1575 he turned, as was natural in a legal family, to Gray's Inn and began to study for the law. The next year, however, he was sent abroad by his father in the company of the English ambassador to France.

For nearly three years Bacon experienced not only events at the foreign court of Henry III but the enormous stimulus of a society in which letters were honored. It was the time of the French Renaissance. Montaigne was being read, the great poet Pierre de Ronsard was writing his autumnal verses. It is likely that in this morally corrupt but brilliant society Francis Bacon received much of the literary stimulus that was to haunt his strangely divided career in the less literate circle of the English court.

In the winter of 1579 Francis had a strange dream. He saw his father's house "plastered all over with black mortar." The dream was prophetic: three days later Lord Keeper Nicholas Bacon was dead.

Francis, the youngest of the several children of two marriages, received only a pittance from the divided estate. His father's unexpected death had left him unprovided for. The student, the man of dreams, was never again to be totally free of financial insecurity. By birth and training he had been fitted for a life far beyond his financial station. And now, where others could be independent, he must humble himself to petition. He was doomed to strive impatiently after what others, securely entrenched, might confidently expect the stream of time to bring to them.

Yet, seen in the afterlight of history, perhaps his misfortune contained a secret blessing. Until the near close of his career he possessed no estates worth confiscating, nothing likely to strike the covetous eyes of kings or their favorites. His chief danger—in spite of his recognition of the neces-

sity of going masked—would be a certain stiff-necked, aris-
tocratic pride showing through in unexplainable odd mo-
ments, in books, in Parliament.

"I have taken all knowledge to be my province," he
once wrote, in importuning aid from his uncle, Lord
Burghley, the Secretary of State. Burghley, that sturdy, im-
perturbable minister, must have shuddered. His nephew
talked as though from a place beyond the century, while at
the same time he revealed a vaulting ambition likely to
cause an experienced courtier to frown. Bacon was eager
and, like all bright youths, occasionally uncertain and
gauche in his impetuous demand for place. Burghley
would give the young man little aid. He was intent on fur-
thering the career of his own son, Robert Cecil, and held
the Bacons—Francis and his brother Anthony—at arm's
length.

All through the reign of Elizabeth—and even though his
cause had been promoted by the Earl of Essex who had
become his patron—Bacon was denied everything but
crumbs. There was one exception: Elizabeth's lease-gift of
the pleasant country residence of Twickenham Lodge in
1595. It was evident that the Cecil faction would never tolerate Bacon's accession to any post of power. Essex's brief
rise to royal favor, and his sudden fall, brought Bacon
nothing but danger and the animus of powerful foes.

All of the more able biographers of Bacon repudiate the
canard that he aided in the fall of Essex and ungratefully
abandoned him. In vain Francis had given Essex sound advice and had pleaded with the frustrated man to abandon
the course which led to his downfall. But the hotheaded,
impetuous courtier, full of the proud violence of a sea

dog's age, had run headlong into revolt and his own death by the axe. Surveying his career today, one might suspect that Essex was not totally sound mentally. Yet he was a brave man and a generous one, and the way his life ended tore many hearts.

Elizabeth herself was never the same afterward; for days at a time, it is said, she would brood motionless, or else suddenly drive a sword into tapestries and hangings, as if in fear of lurking assassins. Bacon was ordered to draw up a state paper explaining the treason of the earl and his accomplices to the restless populace, who had adored Essex. Painfully he did so. His affection for his old patron made the document too lenient for the queen's taste. Those who accuse Bacon of ingratitude might well examine Elizabeth's sharp reaction to Bacon's document: "It is my Lord of Essex, my Lord of Essex on every page; you can't forget your old respect for the traitor; strike it out; make it Essex. . . ."

Until the death of Elizabeth, Bacon's life was largely lived in the shadows, where he wrote state political tracts and advised the queen as a "Counsel Extraordinary." One of his biographers, John Nichol, has commented that "no man ever proposed to enter upon public life with more reason to expect rapid advancement, and very few have had to wait longer for it." The reason is quite plain: in a mercenary age he lacked the means to buy advancement, and the only faction which could have helped him—that of Burghley and, later, of Burghley's son Robert Cecil—regarded him with distaste. His very brilliance was anathema to them.

As for Elizabeth, Bacon once remarked, not unfondly,

to her successor, James I, "My good old Mistress was wont to call me her watch-candle, because it pleased her to say I did continually burn and yet she suffered me to waste almost to nothing." At another time he had spoken bitterly of the way in which he was forced to trudge after small favors like a child pursuing a pretty bird that hops away. Elizabeth could be parsimonious and fickle. As long as Robert Cecil, Bacon's secret enemy, lived, the shadow of that cunning little hunchback would fall across Bacon's path. And when that shadow lifted, Bacon's last great elevation spelled his doom even as he walked into the full sunlight of eminence. He learned what it was to hold power without personal wealth under a king without greatness.

Elizabeth died in 1603 leaving no heirs, or at least none acknowledged. In her final illness, men—swept like sea birds before a rising storm—had posted hard up the great north way to the court of James in Edinburgh: by remote lines of descent he was now heir to the English throne; courtiers had to look to their fortunes. James was not Elizabeth. Essex had favored him. The wheel was turning. "By the mutability of fortune and favor," as the old Elizabethan documents would put it, Bacon's hour would seem to have struck. The Scottish king was reputed to be a learned man. Bacon sought his favor.

The first results were inauspicious: the king had not recognized Bacon, and his former office was not renewed; Robert Cecil, by contrast, remained Secretary of State. The new king intended to knight several hundred people, and Francis—after some correspondence with his cousin Cecil—was one of those selected. But the honor, done hast-

ily to a large body of people, was a small one. A little pen-
sion was given him in remembrance of his late brother
Anthony's services as an intelligence messenger between
Essex and James. He was made a "King's Counsel." The
term meant little: there was a plethora of counselors.
There is evidence that Bacon was in despair and contemplated withdrawing from court life to become a recluse
and scholar.

It is here, while Bacon's star seems waning, that we may
seize the opportunity to examine the intellectual life of
this man who was so long entangled with the practical
affairs of state. As always, a babble of conflicting voices as-
sails us. Some insist that his scholarly abilities were such
that he should never have stooped to politics; others maintain that his talents as a statesman held off the Puritan revolution for a generation and might have prevented it
altogether, had those he served taken his advice.

One thing seems clear: Bacon himself, so far as we are
allowed to penetrate that aloof mask, preferred the clois-
ter. He had, however, one trait which would never have
suited the life of a recluse: he was a man of action. By family and tradition he had been bred to serve the state. Moreover, he was a reformer more than he was a philosopher.
Reserved and shy though he appeared, he was eloquent. In
Parliament he could sway men. He was honored there. Cu-
riously, it was this, and neither his vast learning nor his
dreams, that finally caused James's eye to fall upon him.

In the meantime, however, in the period between Elizabeth's death and the aroused interest of James, Bacon had
little to do. He turned vigorously to the completion of a
book—a book destined to be one of the great books of all

time, even though it was finished in haste in the hope of interesting the learned James. By no means Bacon's first venture into scholarship (his equally well-known volume of *Essays* having appeared in 1597), *The Advancement of Learning* contains the essence of his inner life and his long-frustrated hopes for man.

It is incredible, now, to realize that this great statesman of science was sneered at as a fool by many of his literate fellows in law, government, and the universities. In despair, he had all his works put into Latin because, in that barbaric time, he feared that the rapidly altering English tongue would not survive. Time-conscious as no other man of his era, he viewed books as boats with precious cargoes launched on the great sea of time. One can catch the quality of this time sense, as deep and brooding as that of a modern archaeologist, when he writes:

"But howsoever the works of wisdom are among human things the most excellent, yet they too have their periods and closes. For so it is that after kingdoms and commonwealths have flourished for a time, there arise perturbations and seditions and wars; amid the disturbances of which, first the laws are put to silence, and then men return to the depraved conditions of their nature, and desolation is seen in the fields and cities. And if such troubles last, it is not long before letters also and philosophy are so torn in pieces that no traces of them can be found but a few fragments, scattered here and there like planks from a shipwreck; and then a season of barbarism sets in, the waters of Helicon being sunk under the ground, until, according to the appointed vicissitude of things, they break out and issue forth again, perhaps among other nations and not in the places where they were before."

Through all his trials Bacon's faith in his books, even as lost and bobbing "planks" in the wreckage of time, never faltered. "I have lost much time with this age," he wrote a friend as if, from some high place, his eye spanned centuries. And though he regarded Latin as a more certain medium for survival, it is also known that he promoted the English translation of works that might encourage learning.

We of today have difficulty in realizing that the world of Bacon and Shakespeare was only semiliterate, steeped in religious contention, with its gaze turned backward in wonder upon the Greco-Roman past. Oswald Spengler justly remarks that human choice is only possible within the limitations and idea-forms of a given age. More than three hundred years ago, Francis Bacon would have understood him. Bacon's world horribly constricted his ability to exert his will upon it. At the same time he would have had a slight reservation. "Send out your little book upon the waters," he would have countered, "and hope. Your will may be worked beyond you in another and more favorable age."

For a man whose personal life had been disappointing, Bacon was singularly sure of his destiny. All that he wrote of it has come to pass. The men who destroyed him are remembered, if at all, only because of their perfidious roles in the life of a man whose name now stands with Shakespeare's as the light of the Elizabethan Age.

Other men of Bacon's period were beginning to grope with the tools of science. Only he, however, would clearly

perceive its role and the changes and dangers it would introduce into the life of man. In the years left to him, and
particularly after his fall from office in 1621, a flood of
works poured from his pen. It was almost as if he foresaw
that this would be his last chance to speak "to the next
ages."

There is no doubt that his concentration upon philosophy contributed to his political downfall. It closed his eyes
to signs of public danger; it closed his ears to the machina-
tions of his enemies. His single-minded devotion to duty,
his curious ebullience of temperament, would make him
the easy victim of a political ambush. Nevertheless, the
forces that brought about Bacon's fall might well have
achieved their purpose even against a more unscrupulous
and cunning man.

The times were running against the king, and Bacon was
expendable. In a weird way he would be trapped in a por-
tion of his own political philosophy. But of that, more
later. Here let us examine the essence of that remarkable
book, *The Advancement of Learning,* along with those
works which bear upon his significance as a harbinger of
the modern age.

"This is the foundation of all," wrote Bacon in his
masterpiece, "for we are not to imagine or suppose, but to
discover, what nature does or may be made to do." Bacon's
gift for condensation is so remarkable that it is easy, three
hundred years later, in a different intellectual climate, to
overlook the significance of his words.

The remark just quoted lies at the root of the modern
scientific method. Distilled into one brief phrase, it is the
very essence of science as we know it today. We would

search unavailingly among the practical experimenters of Bacon's time—even the greatest of them—for any comparable analysis of science throughout the whole range of its activity. In a breath he had chained the imagination to reality, but at the same time had left it free to explore the dark crooks and crannies of nature.

We must enter into the intellectual life of Bacon's period if we are fully to grasp the enormity of the task that confronted him, or his challenge to his epoch. I have said that science does not come easily to men; they must be made to envision its possibilities. This was Bacon's role, and it is sheer folly to dismiss him, as some have sought to do, because he personally made no inventions. He did far more; by eloquence and an unparalleled glimpse of the possibilities contained in the new learning, he forced a backward-oriented culture to contemplate its own future.

The magnitude of his educational vision can be perceived only when we realize that well toward the close of the nineteenth century the greatest universities in England were still primarily devoted to the classical education of gentlemen. This fact is both a measure of Bacon's perception and a revelation of the glacial slowness with which ancient institutions are modified. Most of Britain's great scientific contributions in the post-Baconian years had come from members of The Royal Society, an association of scholars which was largely the result of his posthumous stimulus, or from other enlightened amateurs working alone.

"I say without any imposture," wrote Bacon, "that I . . . frail in health, involved in civil studies, coming to the obscurest of all subjects without guide or light, have done enough, if I have constructed the machine itself and the fabric, though I may not have employed or moved it. Before examining the forces that shaped Bacon's thought, let us now consider his mysterious "engine" and see wherein his originality can be said to lie. This is always more difficult to do in the case of a great philosophical thinker than in the relatively simple case of a man who produces a new mechanical invention.

The thinker may range over a wide area, he may reshape old ideas into new and original forms, he may postulate views that are only assimilated or proved long after he is gone. In the end, later comers may either be ignorant of the source of their own thinking or loath to accord to a long-vanished individual credit which might detract from their own originality.

We have to face the fact that the world of scholarship is sometimes a contentious and prickly one. By and large, as the mass of knowledge grows, men devote little attention to the dead. Yet it is the dead who are frequently our pathfinders, and we walk all unconsciously along the roads they have chosen for us. We find what they warned us to look for, and sometimes, also, we are unknowingly entrapped in some half-enchanted circle of ideas woven by a vanished mind. It is a credit to Bacon's perception that, at the very dawn of science, he warned the scholar against this kind of bewitchment, of which the later history of science can provide many instructive examples.

Bacon has long been known as an advocate of inductive reasoning. Indeed, this is a substantial part of his engine for the discovery of the truths of the natural world, the secondary causes which he believed to control all the phe-

nomena of nature. Bacon's emphasis upon induction—
that type of logical thinking by which one ascends from
specific, observed facts to the establishment of general laws
or principles—need not be regarded as original with him,
since the classical world was not unaware of the distinc-
tions between inductive and deductive logic. Bacon, in
fact, never claimed such originality. What he did seek to
do with his new *use* of induction was to avoid the sterile
logic of the Aristotelian schoolmen. Since this type of
thought has practically vanished from the modern world,
we forget that education, in Bacon's day, was largely
confined to metaphysical argument along with the reading of Greek and Roman classics. The techniques of logic,
in other words, were being expended upon abstract controversy, while nature itself passed largely unexamined.

Men, to paraphrase Bacon, were spinning webs out of
their own substance. To recapture reality it would be necessary to bring speculation into conformity with reality, to
ascend from genuine facts to deductions, and to avoid
hasty and unsubstantiated theory. As one student of the
time has remarked, people "decided all questions not by
investigating the observable facts, but by appealing to the
infallible authority of Aristotle." Around the scholars of
Elizabeth's century lay a natural universe scarcely investi-
gated except for the exploration feats of the often unlet-
tered voyagers.

Bacon was convinced that once man came to understand
this unexplored nature about him, he could attain power
over it, but that this potential power could be achieved
only by the right methods of investigation exerted on a
very large scale. "Looking back," says the English philoso-

pher C. D. Broad, "we can see that he was right, and we may be tempted to think that it was obvious. But it was not in the least obvious at the time; it was, on the contrary, a most remarkable feat of insight and an act of rational faith in the face of present appearances and past experience."

Bacon's associates in government, including Elizabeth and James, had been brought up in the traditional learning. James, in particular, prided himself on an antique, pretentious classicism. Neither was impressed by so unconventional an idea as Bacon's and one which, if adopted, would upset the prevailing school system. Both rulers were entrapped in a maze out of which he was powerless to lead them. The more he wrote in the vein of his own convictions, the more warily he was regarded by his political contemporaries.

Experimenters—such as William Harvey, the discoverer of the circulation of the blood—were beginning to appear, but none seems to have had Bacon's total vision of what science and the experimental method could achieve over the centuries. Harvey, in fact, referred amusedly to Bacon as writing of science "like a lord chancellor." His remark is true, but imperceptive. Bacon was the first great statesman of science. He saw its potentiality in the schools; he saw the necessity of multiplying researchers, establishing the continuity of the scientific tradition, and promoting government-supported research for those studies which lay beyond private means and which could not be accomplished "in the hourglass of one man's life."

This vast vision could only have emerged from a mind trained to state affairs, to the management of kingdoms, and withal, a mind equally devoted to discovery. It is ridiculous to bemoan Bacon's practical experience of statecraft; it contributed enormously to his insight. The pity lies in the fact that he came so close to the seats of power without the opportunity to realize his dreams. It is an apt illustration of the degree to which even a great genius can be restricted and made helpless by his time. Yet in justice we must add, not totally so. The prestige of Bacon's final offices gave greater weight to his literary pronouncements, financed his publications, and in other indirect ways, lent wings to his words beyond what would have been possible for an obscure scholar opposed by many of his compatriots.

Another aspect of Bacon's contributions which deserves attention is his conception of the *mundus alter*—"the other world" produced by human culture, a world drawn out of the void and made possible by the arts of man. It is, in a sense, a latent world filled with novelties which man, by his own ingenuity, can bring out of nature. Until the time of Bacon, man had more or less "drifted" in the natural world. His culture, with all its rational and irrational elements, had grown up largely without conscious self-examination or attention to the fact that man might possess the power successfully to mold and improve his own society through science.

Bacon, by contrast, was intent to turn man into an actively anticipatory creature rather than a backward-yearning one. In doing this he contended against great obstacles. He fought against the vested interest of the Scholastic teachers indifferent to experiment; he inveighed against the widespread belief that the classical past would never be equaled because the world was far sunk in decay and destined to perish at no very distant date. This last notion, which was widely accepted and promoted, was destructive of initiative and conducive to indifference.

Bacon struggled against this despair with every means at his command. "If we must select some one philosopher as the hero of the revolution in scientific method," said William Whewell, a learned nineteenth-century historian of science, "beyond all doubt Francis Bacon occupies the place of honor." He based his estimate upon Bacon's conception of the dawn of a new era and the shifting of logic from contention to its use in the analysis of experiments. "In catching sight of this principle, and in ascribing to it its due importance," continues Whewell, "Bacon's sagacity wrought unassisted and unequalled." Ungrounded argument, Bacon saw, must be replaced by a logic applied to reality. Only then could man bring his second world into being.

Considering the time at which he wrote, Bacon's numerous insights are phenomenal. For example, although he himself was not a mathematician, he foresaw the necessity for using mathematics in some of the more subtle examinations of nature. The mathematician A. R. Forsyth dwelt upon this aspect of Bacon's thought before the British Association for the Advancement of Science in 1905, and it has not gone unnoted by others.

Similarly, Bacon reveals perceptive insights into biology and anthropology. He raised, in a quite objective way, the question of whether the transmutation of species could occur and commented that the problem required deep research. He refused the arbitrary abandonment of the idea.

He observed that the lower organisms might reveal secrets of life which in the higher organisms "lye more hidden."

His analyses of the "cave of custom" and of the necessity for understanding the "Idols" that distort the thinking of the average man are the product of long observation of men under emotional stress. Jonathan Wright, a number of years ago, commented that it is to Bacon that we owe the idea of utilization of controls in scientific experiment.

One could point to many other evidences of Bacon's wide-ranging mind—his early recognition of the value of the history of science, his contention that biography should not be confined to rulers, and last, but not least, his recognition of the value of the division of labor in science. Above all else he dwelt upon science "for the uses of life." He warned that knowledge without charity could be as dangerous as the modern world has finally discovered it to be. In contrast to today's warring nationalisms, Bacon spoke in *The Great Instauration* of bearing a strong love for "the *human republic,* our common country."

If one were to ask why science arose when it did, and why at its dawn so great a spokesman should have appeared to spread its doctrines, one would have to pursue innumerable beams of light into that globe of crystal which Bacon termed the human understanding. His own mind serves as a kind of condensing lens in this respect. Analogies in language drawn from the great voyagers dot his pages. It is evident that the geographical discoveries of his time, and the circumnavigation of the earth, had set wise men's statements at nought and promoted the independent examination of the natural world. The historian Merle Curti, commenting upon his influence in pioneer

America, remarks, "It was no accident that Francis Bacon's ideas were rediscovered and put to work in an era characterized by the rise of the common man."

Formal theology had been shaken by observations in continents unmentioned in the Bible. A second Book of Revelation, the book regarded as unclouded by human error and confusion—the book of Nature—was becoming increasingly respectable to devout minds. Bacon's mother was under strong Puritan influence. The Puritan desire to rebuild an earthly Paradise, even in the wilds of the New World, was growing. It is no secret that later, in New England, the Puritan clergy promoted the new astronomical discoveries. This is not to equate science with Puritanism alone, but it does suggest that something about the Reformation played a part in the emergence of a full-bodied scientific movement which, a generation or so after Bacon, recognized his significance as the great spokesman for the scientific method itself and all that was to follow in its train.

No single man, of course, can be credited with the creation of modern science. A unique institution had appeared, however, with an equally unique spokesman. One can only repeat that no other figure from that dim light of the scientific morning catches more powerfully at our imagination than the man who wrote: "I could not be true and constant to the argument I handle, if I were not willing to go beyond others; but yet not more willing than to have others go beyond me again."

Science, he maintained, was "not a belief to be held but a work to be done." He thought of it as "a work for the ages and the peoples. . . ." In that work and its promulgation he had worn out such days as were given him. If he was marked, in any small degree, by the times he lived in, who are we of the century of My-lai and the mushroom cloud to sit in unrelenting judgment?

We left Bacon's political career at a time of crisis, but with a hint of what was to come. He had desperately hurried forward the completion of *The Advancement of Learning,* in order to impress the new king James. That monarch, whose backward-directed classical learning would find little it could comprehend in the *Advancement,* was eying quite another aspect of Bacon's diversified career: his role in Parliament. Bacon was highly respected there.

Bacon had sat in the Commons since the days of Elizabeth; in fact, he had once aroused her fury by opposing a tax requisition which he felt was inordinate and would cause suffering among the poor. For those who conceive of Bacon as an unscrupulous manipulator for favor, his letter to Lord Keeper Puckering on that occasion is worth quoting: "It mought please her sacred Majesty to think what my end should be in those speeches, if it were not duty and duty alone. I am not so simple but I know the common beaten way to please."

In 1603 the problems of Parliament versus the king were to emerge once more in another guise. There were church problems; there was the problem of the political union between England and Scotland. James was a newcomer from a more absolutistic and barbaric land. He needed advice and a trusted statesman who could mediate between the Lords and Commons. Bacon's moderation, his ability to sway audiences, the respect accorded him in the lower house, all fitted the needs of the new regime. It was a time when, in the words of Fulton Anderson, one of the most careful students of Bacon's career: "A wise king would decrease the area of his prerogative and gradually increase the range of his subjects' privileges. For some sixteen years," Anderson records, "Bacon would be trying to make James aware of these things; but the King would prove neither wise nor teachable."

Bacon assumed his first important administrative post under James as Solicitor General in 1607. It was an exacting office, made more so by Bacon's diverse abilities as a statesman. Frequently he was engaged in carrying out duties that ordinarily would have been assigned to others. For a man of frail health his energies seem almost superhuman. In the midst of difficult affairs of statecraft he still yearns over his "rebirth" of the sciences, and studies ways that Cambridge or Oxford might be encouraged toward the new learning—toward laboratories, "engines, vaults, and furnaces."

All this time the king's treasury, in the hands of Robert Cecil, was sinking into debt. Taxes and impositions levied upon the public were growing more onerous. Opposition to James and the highhandedness of his favorites was growing in the House of Commons. These events were the first weather signs that were to lead to the Puritan revolution. Bacon could read them, but those around him could not. Blind though he might sometimes be to his personal interests, he was never blind to the interests of the state. A long list of state papers and unheeded advice testifies to his efforts.

He would have been an able replacement as Secretary of State for Robert Cecil, who died in 1612 leaving James's treasury empty. Bacon sought the post, but James, wary of Bacon's association with the now-feared Parliament, turned aside from the one logical candidate. He named instead an inexperienced man with no obligations to the House; Bacon was appointed Attorney General. He would yet be made Lord Keeper in 1617 and rise to the high office of Lord Chancellor of England in 1618.

Meanwhile his new office of Attorney General, as events were to prove, was a far more vulnerable post than his previous one. Ironically, the suspicious James, in spite of not trusting Bacon sufficiently to allow him to succeed Cecil, continued to seek his advice on the financial affairs of the kingdom. Once more, in the words of Anderson, "the courts and the constitution were to be preserved in a continuity through great trials and hazards by one man and one man alone, Francis Bacon. In this regard he became for a period the chief axial officer of the kingdom. He managed, without ill-deserving, to keep the constitution intact . . . to maintain the law and liberty of subjects, and to preserve 'the King's honor,' through wise, skillful, and just resorts."

Only in the reign of James's son Charles I would the great storm of the Puritan revolution finally strike. To Bacon is owed no small credit for delaying that storm by a generation. If James had earlier given heed to Bacon's counsels, if Buckingham and Cecil had been less venal, the interruption of the British monarchy and the sorry death of Charles need never have occurred at all. A people's great tragedy was winding to its conclusion—a tragedy which would not end with Bacon's death but only when James's son would lay his head upon the block in payment for his own and his father's obstinacy.

As part of his monarchial creed, Bacon had once written: "It is well, when nobles are not too great for sovereignty nor for justice, and yet maintained in that height, *as the insolency of inferiors may be broken upon them before it come too fast upon the majesty of kings*" (italics mine). This maxim is practiced even in modern democracies, though generally under less severe conditions. Francis Bacon, the moderate monarchist, Parliament's man, the sensible compromiser, was now to act out in life his own observation.

He had risen to high office. The Lord Chancellorship of England, given to him in 1618, had brought with it the title of Baron Verulam. In the very year of his tragic fall—1621—he was to receive yet another honor: the investiture as Viscount St. Alban. On January 27, five days after his sixtieth birthday, his new dignity was conferred upon him in full ceremony. In a letter to the king, thanking him for this new advancement as well as the previous ones, Bacon added: "And so I may without superstition be buried in St. Alban's habit or vestment."

He was not to die for five years. But he was to be broken by the mob pressing more and more angrily upon the "majesty" of James. High office could not save him. In Parliament a majority of the members, manipulated by such enemies of Bacon's as the sadistic Sir Edward Coke, would turn against him. In frustration at their inability to vent their rage on the king, or on his favorite George Villiers, they would destroy the one man who had sought to temper the royal excesses and preserve the state.

These acts were far from ingenuous. Old enemies would scurry on black errands; traditional homage would be deliberately redescribed as bribery in order to fit the scene; the lapses of servants would be laid upon the overworked master. Southampton, former conspirator with Essex, would charge Bacon with corruption in office. One has only to consider the way the majority of these men were to end, amidst their own violence and rapacity, to realize the nature of Bacon's judges. James wept; Villiers vacillated. "My lord," said Bacon when confronted with the charges, "if this is to be a Chancellor, I think if the Great Seal lay upon Hounslow Heath, nobody would take it up."

James did not dare dismiss Parliament; the unscrupulous Villiers, though obligated to Bacon for advice in earlier years, was not a man of sentiment. We need not linger over the intricacies of things said and done, nor over Bacon's personal agony. He intended his defense, but there came before it a fatal interview with James. James advised him—and to a man in Bacon's position this was a command—to avow his guilt and trust his protection to the Crown.

It was the royal will. When Bacon left the king's presence, he is reputed to have remarked: "I am the first sacrifice; I wish I may be the last." He was face to face with his own dictum that in times of crisis the king's circle was expendable for the protection of the monarch. Without land or title of his own, he had been raised high by James. He owed him much. Bacon bowed to the King's wishes.

He must have felt, as Charles Williams, one of his most perceptive biographers has remarked, that he was "on a ship fantastically laboring on a wild ocean, manned with imbeciles."

For more than three centuries the intellectual world has contended, censured, moralized, and probed into a world that is gone. The man behind the mask gives back no answer, the man who took care with his documents, those frail vessels in which he put his final trust, utters to posterity no word of vituperation or defense. He was a man who lived by his code, a servant to his monarch and the state.

The world was changing. Even that monarch, in spite of his avowals and a few protective gestures, did little to alleviate Bacon's lot or clear his name. Charles I after him did no more. A terrible loneliness descended on the dying man, the last and least publicized of the Elizabethan explorers—the stay-at-home discoverer of the New Atlantis, the opener of the great door into the future.

About him the storms of a new era were gathering. In a generation men would begin to talk of Bacon again, scientific societies would arise on the plans he and other dreamers had modeled. Men would grow curious about him, but the stain would be there, the ineradicable, terrible stain, unexamined in its time or place or condition, spreading endlessly as his name spread, as it still spreads today.

Strangely, the mask dropped just for a moment after his death. Long ago he had courted the Lady Hatton, who had cast him aside for his old rival, Sir Edward Coke, one of the undoubted contrivers of his downfall. She had grown to hate her overbearing, quarrelsome husband; Bacon had been equally unfortunate in his own marriage.

When the sharp winter airs of which he had complained
struck him down at last, there was found in his will an intended gift to Lady Coke. The episode is unexplained to
this day. Perhaps Bacon's legacy was dedicated to some lost
spring by a man who, rumor says, was cold to the ways of
love.

The harsh accusers of his own time have long been silent, but Bacon's ordeal has not ended. What we find in
him now may essentially be a measure of the nature of
ourselves as individuals. "Bacon's wisdom," said the arch-
bishop Richard Whately, "is like the seven-league boots
which would fit the giant or the dwarf, except only that
the dwarf cannot take the same stride in them."

2

A Hidden World

In 1620 THE PILGRIMS LANDED UPON PLYMOUTH Rock. Fifteen years before, just prior to the time the Jamestown colonizers boarded ship, there was published in England a book whose author had been powerfully influenced by those western waters over which a few voyagers were beginning to find their way by primitive compass into a new and hidden world. An early edition of that book—*The Advancement of Learning*—lies before me as I write. Sir Francis Bacon, its author, was a stay-at-home destined all his life to hear the rumble of breakers upon unknown coasts.

The book from my shelves is an edition published in 1640, just fourteen years after Bacon's death. It is crinkled with age and touched by water, and its pages are marked by the rose of a creeping fungus. It has passed through the English civil wars, and Cromwell's cavalry has ridden hard in the night by its resting place. It has been read and dismissed and pondered again by candle, by woodfire, by gaslight. Somewhere in the more than three hundred years of its journey it crossed the western ocean that moved so powerfully the mind of its creator. There are pages so blackened that one thinks inevitably of the slow way that the fires in the brain of genius run on through the centuries, perhaps to culminate in some tremendous illumination or equally unseen catastrophe.

Not all men are fated, like Sir Francis Bacon, to discover an unknown continent, and to find it not in the oceans of this world but in the vaster seas of time. Few men would seek through thirty years of rebuff and cold indifference a compass to lead men toward a green isle invisible to all other eyes. "How much more," he wrote in wisdom, "are letters to be magnified, which as ships pass through the vast seas of time, and make ages so distant to participate of the wisdom, illuminations, and inventions, the one of the other. . . ." "Whosoever shall entertain high and vaporous imaginations," he warned, "instead of a laborious and sober inquiry of truth, shall beget hopes and beliefs of strange and impossible shapes." It is ironic that Bacon, a sober propounder of the experimental method in science—Bacon, who sought so eloquently to give man control of his own destiny—should have contributed, nevertheless, to that world of "impossible shapes" which surrounds us today.

Appropriately there lingers about this solitary time-voyager a shimmering mirage of fable, an atmosphere of mystery, which frequently closes over and obscures the great geniuses of lost or poorly documented centuries. Bacon, who opened for us the doorway of the modern world, is an incomparable inspiration for such myth-making proclivities. Rumors persist that he did not die in the year 1626 but escaped to Holland, that he was the real author of Shakespeare's plays, that he was the unacknowledged son of Queen Elizabeth. Rumor can go no further; it is a measure of this great discoverer's power to captivate the curiosity of men—a power that has grown century by century since his birth in 1561. In spite of certain mystifying aspects of his life, there is no satisfactory evidence sufficient to justify these speculations, though a vast literature betokens their fascination and appeal.

Perhaps we do not really wish to acknowledge that time and misfortune can fall upon men who, to our more humble minds, seem veritable gods. Perhaps mankind has staked in such geniuses its spiritual effort to evade extinction. The present decade, with all its threatened and actual violence, a time of intercontinental missiles, fantastic computers, and words passing us invisibly upon the air, marked the four-hundredth anniversary of the seer who first formulated the vision of our time and who, perhaps more than any other man, set men consciously upon the road of modern science. "Nature," he wrote in *De Augmentis Scientiarum,* "exhibits herself more clearly under the trials and vexations of art [that is, experiment] than when left to herself." Bacon defined, in addition, the image of what a true scientist should be: a man of both compassion and understanding. He whispered into men's careless ears that knowledge without charity could bite with the deadliness of a serpent's venom.

As more and potentially deadlier satellites wing silently overhead, it is plain at last that Francis Bacon's words were true, though his warning has passed unheeded

through the three short centuries of unremitting toil
which have altered the face of the planet and consumed
the green forests of America. Finally men have forced the
gateway of space which the average Elizabethan regarded
as the empyrean realm of deity never to be penetrated by
man.

"I leave my name," reiterated Francis Bacon, "to mine
own countrymen after some little time be passed." These
words were written in the knowledge that his voice had
passed largely unheard in his generation and that, in addition, his calculated fall from royal favor had cost him his
good name. As I have previously noted, in the Tudor century of Elizabeth and the Stuart reign which followed it,
the absolute right of monarchs brought many an innocent man and woman to the headsman's block or to forced,
communist-like confessions of guilt, as state expedience
demanded.

With this kind of atmosphere Francis Bacon was all too
familiar. Condemned to spend his life in a world of sometimes petty, often dangerous court intrigue, he chose to
have his actual being in a far different one of his own making. He sought to encourage a world the monarchs he
served could not envisage, a world of schools, of research
promoted by the state, a world in which more could be
done than that which could be projected in a single generation.

It is ironic that this man, who died broken and forlorn
in an age he never truly inhabited, was warmly loved by
such great literary figures as Ben Jonson but arrogantly
dismissed by William Harvey and other intellectuals. Sir

Edward Coke, the eminent jurist, inscribed in one of Bacon's books the contemptuous comment:

> It deserves not to be read in Schooles
> But to be freighted in the Ship of Fooles.

James the First, the monarch he served, remarked cynically that Bacon's work was like the peace of God that passes all understanding.

King, jurist, scientific experimenter alike saw nothing where Francis Bacon saw a world, a new-found land greater than that reported by the homing voyagers, the Drakes, the Raleighs, of the Elizabethan seas. We of the western continent that stirred his restless imagination should particularly honor him because we are also a part, a hopeful beach mark, of the world he glimpsed in time. His great book *The Advancement of Learning* belongs more truly to us today than it ever belonged to that powerful but imperceptive monarch to whom it was originally addressed. We are fortunate in being allowed to share the ideas and visions of a statesman-philosopher whose influence has transcended, and will continue to transcend, his own century.

Man, Bacon insisted, must examine nature, not the superstitious cobwebs spun in his own brain. He must ascertain the facts about his universe; he must maintain the great continuity and transmission of learning through universities dedicated—not to the dry husks of ancient learning alone—but to research upon the natural world of the present. Then, and only then, would his second world, the invisible world drawn from man's mind, become a

genuine reality. In the *Parasceve* he wrote: "In things artificial nature takes orders from man and works under his
authority; without man, such things would never have
been made. But by the help and ministry of man a new
face of bodies, another universe or theatre of things, comes
into view." Though he sought to combine the discoveries
of the practical craftsmen with the insights of the philosopher, Bacon saw more clearly than any of the other Renaissance writers that the development of the experimental
method itself, the means by which "all things else might
be discovered" was of far more significance than any single
act of invention.

It was Bacon's whole purpose, set against the Scholastic
thinking of medieval times, "to overcome," as he remarks
in another of his works, the *Novum Organum,* "not an ad-
versary in argument, but nature in action." Truth, to the
medieval schoolmen of the theologically oriented universities, rested upon the belief that reality lay in the world of
ideas largely independent of our sense perceptions. In this
domain the use of a clever and sophisticated logic for argument, rather than observation of the phenomena of nature, was the road to wisdom.

In *The Advancement of Learning,* the *Novum Organum,* and several of his other works, Francis Bacon, as
we have seen, presented an "engine," for the attainment of
truth; namely, induction. We must refrain, Bacon contended, from deducing general laws or principles for which
we have no real evidence in nature. Instead, because of our
human tendency to leap to unwarranted conclusions, we
must dismiss much of what we think we know and begin
anew patiently to collect facts from nature, never straying

far from reality until it is possible through surety of observation to deduce from our observations more general laws.

It can, of course, be argued that without some hint or idea of what we seek, fact-gathering will take us nowhere. Bacon, however, was contending against a philosophy almost diametrically opposed to the examination of the natural world. He himself was aware of the way the mind runs in ascending and descending order from fact to generalization and back again, "from experiments to the invention of causes, and descending from causes to the invention of new experiments." "Knowledge," Bacon insists specifically in *The Advancement of Learning,* "drawn freshly and in our view out of particulars, knoweth the best way to particulars again. And it hath much greater life for practise when the discourse attendeth upon the example, than when the example attendeth upon the discourse." "We have," he emphasizes, "made too untimely a departure and too remote a recess from particulars"—that is, from the facts of the natural world about us. Indeed, it was just such facts, impossible to be subjectively gathered from the mind alone—fossil bones, superimposed strata, animal and plant distributions—which led, more than two hundred years after Bacon's death, to perhaps the greatest inductive achievement of science—the creation of the evolutionary hypothesis.

Yet Bacon, for all his emphasis on observation, was ahead of his time and writes, indeed, like a modern theoretical physicist when he argues that "many parts of nature can neither be invented—that is, observed—with sufficient subtlety, nor demonstrated with sufficient perspicuity . . . without the aid and intervening of the mathematics."

Bacon was certainly not unaware of the intuitive insights of genius or the fact that time might yield new ways of discovery. He saw, however—and this has been taken as an affront by the sensitive—that reliance upon the sporadic appearance of genius was no sure or sane road into the future. Long before the rise of sociology and anthropology, he had grasped the concept of the cumulative basis of culture and the fact that inventions multiplied in a favorable social environment. Science and its traditions had to be transmitted through the universities, and its efforts had to be publicly supported. By such means humanity could progress more consistently and far better sustain its great intellects than if such intellects were allowed to flounder amidst an unenlightened and hostile populace. Indeed with laboratory research conducted on a high level there would be less distinction between the achievements of able teams of researchers and the occasional attainments of unsupported genius under the handicaps so evident in Bacon's day.

Bacon's observations were perfectly justified. Our entire school system, however faulty it may remain in particulars, is predicated upon Bacon's faith in the transmission of learning and the continuing expansion of research into that dark realm where man, either for human benefit or terror, can increasingly draw the purely latent, the *possible,* out of nature, thus supporting Bacon's conception that the investigation of nature should include a division "of nature altered or wrought." What the philosophers of a later century would call the emergence of novelty in the universe was already clear in his mind.

Experiments of light, Bacon insisted, were more important than experiments of fruit. He sought to restore to man something of his fallen dignity, to erase from his mind the false idols of the market place and to regain, by patient labor and research, some remnants of the innocent wisdom of Adam in man's first Paradise. The mind, he contended, had in it imaginative gifts superior to the realities of sixteenth-century life; in fact, to the realities of the world we know today.

Our ethics are diluted by superstition, our lives by self-created anxieties. Our visions have yet to equal some of his nobler glimpses of a future beyond our material world of easy transport, refrigeration, and rocketry. The new-found land Bacon sighted was not something to be won in a generation or by machines alone. It would have to be drawn slowly, by infinite and continuing effort, out of minds whose dreams must rise superior to the existing world and shape that world by understanding of its laws into something more consistent with man's better nature. "Our persons," he observes, "live in the view of heaven, yet our spirits are included in the caves of our own complexions and customs which minister unto us infinite errors and vain opinions if they be not recalled to examination."

Bacon's science, as he formulated it, was to have taken account of man himself and to have studied particularly the ethical heights to which individuals, and perhaps after them the mass, might attain. His "second world," as his late fragment, the *New Atlantis*, indicates, was to be a world of men transformed, not merely men as we know them amidst the machines and pollution of the twentieth century. Thus parts of Bacon's dream lie beyond us still.

The man who opens for the first time a doorway into the future and who hears faint and far off, like surf on unknown reefs, the tumult and magnificence of an age beyond his own is confronted not alone with the scorn of his less perceptive fellows, but even with the problem of finding the words to impose his vision upon contemporaries inclined to the belief that the world's time is short and its substance far sunk in decay. To achieve this well-nigh impossible task, Bacon had to take the language of his period and, like the seer he was, give old words new grandeur and significance, blow, in effect, a trumpet against time, darkness, and the failure of all things human.

It was his task to summon the wise men, not for one day's meeting or contention, not to build a philosophy of permanence under whose shadow small men might sit and argue, but to leave, instead, a philosophy forever and deliberately unfinished. His philosophy, for human good or ill, has brought the foreshore of a great and unknown continent before our gaze. On the beaches of this haunted domain we find the footprints of the greatest Elizabethan voyager of all time—a man who sounded the cavernous surges of the darkest sea against which men will ever contend: the sea of time itself.

It is not possible to realize the full magnitude of Bacon's achievement without some knowledge of this age of the scientific twilight—an age when men first fumbled with the instruments of science, yet in the next breath might consider the influence of stars upon their destinies or hearken to the spells of witchcraft. It was generally held that the human world was far gone in decay. Men who had only lately tasted the fruits of the lost classic learning had their gaze cast backward upon a civilization they felt unworthy to equal. The giant intellects of Greece, the vanished splendor of the Roman empire held the human mind enthralled; the voices of antiquity ruled the present. This spell—for so it seemed to lie upon men's minds—was the result of a belief in a world whose destiny was incredibly short by the scientific standards of today.

Man's fall from the perfect Garden was believed to have infected nature itself. The microcosm, man, had destroyed the macrocosm, nature. The human drama of the Fall and the Redemption was being played out upon the brief stage of a few trifling millennia. Eternity, the timeless eternity of the spiritual world, was near at hand. By contrast, the long story of geological change and evolution was unknown. The smell of an autumnal decay pervaded the entire Elizabethan world. Over all that age which now glimmers before us with jewels of thought more wonderful than the gems from some deep-buried pirates' hoard, there was a subdued feeling in men's hearts that the sands in the hourglass were well-nigh run. It was autumn, late autumn, and God was weary of the play.

Men had come a long way down from the ages of the patriarchs recorded in the Bible, or from those philosophies which had been spun under the younger sunlight of Greece. Man was truly caught up in a spell of his own making; the contentions of past philosophers were on his lips; their systems occupied his brain. Bacon, by contrast, saw the wisdom of the ancients "but as the dawning or break of day." "I cannot," he emphasized in *The Advancement,* "but be raised to this persuasion, that this third period of time will far surpass that of the Grecian and Roman learning: only if men will know their own strength and their own weakness both; and take one from the other light of invention, and not fire of contradiction." "Employ wit and magnificence to things of worth," he chided, "not to things vulgar."

Elizabethan craftsmen or lonely seafarers sometimes made discoveries, he noted, but the experiments of calloused hands were often scorned by gentlemen. After all, what did newcomers matter in a world of falling leaves, with even the moon grown old and spots upon its face? In a sense these people lived as we today live under the shadow of atomic disaster, with a foreboding that the human future is running out. We know a universe of greater antiquity, but we, in a different way from the Elizabethans, know our own sinfulness with a dreadful renewed certainty. It is with us as they thought it was with them. Our civilization smells of autumn. "Lest darkness come," St. John records the remarks of Jesus in the New Testament, "believe in the light, that ye may be the children of light."

Francis Bacon never wavered in the faith that this light was compounded of both knowledge and charity. With the first of these virtues he sought to open up a hidden continent for man's exploitation. Of his second virtue, it is all too evident that it remains today largely disregarded. Yet only by charity and pity did Bacon foresee that man might become fit to rule the kingdom of nature. The technologi-

cal arts alone had concealed in them, he realized, a de-
monic element. They could bring men riches, but they
could draw out of nature powers which then became non-
natural because they were subject to the human will with
all its dangerous implications.

Bacon was the unwilling servant of an age he loathed.
Yet he has been charged with calling up from the deeps of
time that which we, in the modern world, have not found
the power to put down. We cannot say Bacon did not
warn us. The contemplation of the light of wisdom, he argued, was fairer than all its uses. He pleaded for the in-
struction of youth in high example. His warning to begin
at the very threshold of his new continent the search for
protection against its dangers went unregarded.

Bacon knew and said repeatedly that the light of truth
could pass without harm over corruption and pierce un-
sullied the darkest and most noisome sewers. Perhaps in
this sense he foresaw how far into the depths of ourselves
we must descend in order to vanquish that serpent of evil
which, projected into biblical mythology is, in fang,
venom, and scale, a considerable part of our long evolutionary heritage and, not incredibly, perhaps, our neme-
sis. Yet if Bacon's contemporaries felt the blast of some
final autumn at their backs, Bacon, the midnight reader
and stay-at-home, had sensed and transferred to his own
peculiar dimension a fresher and more vital wind.

It was something the giants of antiquity had not
mastered—the wind of the world-girdling voyagers. "This
proficience in navigation and discoveries," he wrote,
"may plant also an expectation of the proficience and aug-
mentation of all sciences: because it may seem they are or-

dained by God to be coevals, that is, to meet in one age."
Here in this supposed dead season of the intellect existed
a triumph unclaimed by antiquity. Not for nothing did
the first edition of his *Great Instauration* show a ship passing in full sail through the pillars of Hercules; not unknowingly did Bacon speak of himself as a stranger in his
century—an explorer passing as warily as a spy through
foreign lands.

The voyagers had brought home tales of continents and
men undreamed of within the little confines of Christian
Europe. Bacon, sensing almost preternaturally the meaning of that oceanic wind, had raised a moistened finger in
what, by contrast, seemed the dead calm of a few sparse
human generations. Against that finger, though faintly, he
had sensed the far-off wind of the future. He had envisioned man's power to change and determine his own destiny. Scientifically, he was one of the first to grasp the latent novelty that could be drawn out of nature. He was
beginning to discover history and world time—a phenomenon which the historian Friedrich Meinecke was later to
call "one of the greatest spiritual revolutions which western thought has experienced." It meant, as Lord Morley
observed astutely, "the posing of an entirely new set of
questions to mankind."

In that far continent in time, against whose coasts
Bacon caught the murmur of troubled surf, "many things
are reserved which kings with their treasures cannot buy,
nor with their force command, their spials and intelligen-
cers can give no news of them, their seamen and discoverers cannot sail where they grow." This hidden world, he
argued, could be brought out of nature only by a great act

of the human imagination. "Reason," he proposed, "beholds a farre off even that which is future." There is possible, he insists, a kind of natural divination, a key to the opening of nature's secrets, rather than the idle acceptance of immobility, of pure dogmatism, of "animal time." The sovereignty of man over nature, Bacon contended, lies in knowledge. "For that which in these perceptions appeareth early, in the great Effects cometh long after." "It is true also," he wrote, "that it serveth to discover that which is hid, as well as to foretell that which is to come. . . ."

What Bacon lacked as an experimenter he made up for in his range and vision of what science in its totality meant for man. Bacon was an experienced planner and adviser upon matters of diplomacy and affairs of state. He was, in fact, the first scientific scholar to approach the incipient institution of science from the viewpoint of a practical statesman. As he so vividly remarks, "though the whole earth were nothing but . . . academies of learned men, yet without such an experimental history as I am going to describe, no progress worthy of the human race . . . could be made. Either this must be done, therefore, or the business must be abandoned."

The individual discoverers would come later. This, with sure instinct, he knew. The real problem was to break with the dead hand of the traditional past, to free latent intellectual talent, to arrest and touch with hope the popular mind, to carry word of that which lay beyond the scope of the isolated individual thinker; namely, to dramatize what we have previously referred to as the invention of the experimental method itself—the invention of inventions, the door to man's control of his own future. "Every act of discovery," Bacon wisely enjoins us, "advances the art of discovery."

Bacon's discoveries lie in the intangible realm of thought. "Deeds need time," wrote Friedrich Nietzsche, "even after they are done, to be seen and heard." The mind of Bacon, whose thoughts are still moving toward and beyond us across the dark void of centuries, already shines with the impersonal radiance of a star.

What we make of Bacon's second world in every human generation lies partly in the unfathomable realm of human nature itself. We may dip into his pages and write of him as one who, unwittingly, set us on the road which led at last to stalemate by terror, to a world divided by a fence of mine fields, barbed wire, and concrete blocks. Just as readily we could establish that he urged from the beginning of science that man's own powers must be used with wisdom. The poet Samuel Taylor Coleridge, a great student of Bacon, was actually echoing the Lord Chancellor when he said that man was a secondary creator of himself and of his own happiness or misery.

In *The Advancement of Learning* Bacon makes clear his concern, not only with knowledge, but its application for human benefit and freedom. He knew that man himself, unless well studied and informed, was part of the darker aspect of that unknown country which, as he said, "awaited its birth in time." "Mere power and mere knowledge exalt human nature but do not bless it," he insisted. "We must gather from the whole store of things such as make most for the uses of life."

For the uses of life! I switch off my reading light for a

moment and a knob manipulated by my hand brings from the ends of the earth threat and counterthreat. The wars of the Cavaliers are ended. Men speak to each other now like the wrathful God of the Old Testament, threatening to make of their enemies' countries dust and a habitation for owls. The threats are real; the power, torn from nature, lies exposed in human hands. The voices pass, faintly contending, on their way to the vast silences of space.

For the uses of life. I repeat Bacon's phrase in pain, in the darkness of my study. We who are small, and of those unknown generations for whom Bacon labored and to whom he left his name for judgment, are now ourselves to be judged. Charles Williams, one of Bacon's biographers, has remarked perceptively that Bacon's strange and shadowed life, his proffer of powers from which men shied with reluctance, was the heaviest burden of his genius and his rejected love for mankind. "He brought with him," confessed Williams, in the days before the atomic era, "something that might easily become a terror. Men like Bacon are not easily loved or used: something terrific exists in them, however humbly they speak." "Even to deliver and explain what I bring forward," Bacon once remarked in weariness, "is no easy matter, for things in themselves new will yet be apprehended with reference to what is old." In the passage of long centuries the endless innovations of science have not quieted that lust for power which still blocks the doorway to the continent of Bacon's dreams.

Written into *The Advancement of Learning* in words that should be read and reread by every man who calls himself civilized is Bacon's key to the true continent. It has always been known to the great seers, but now, and all too rapidly, it must become the property of mankind in general, or mankind will perish. "The unlearned man," wrote Bacon carefully, "knows not what it is to descend into himself, or call himself to account . . . whereas with the learned man it fares otherwise, that he doth ever intermix the correction and amendment of his mind with the use and employment thereof."

"For the uses of life," we might well reiterate, for so he intended, and this is why his green continent lies beyond us still in time. Five little words have shut us, with all our knowledge, from its shores—five words uttered by a man in the dawn of science, and by us overlooked or forgotten.

Francis Bacon coughed out his life alone, in a century to which he had been a stranger. He had once written to a friend, "I have lost much time with this age; I would be glad to recover it with posterity." Bacon saw, like the statesman he was charged with being, the full implications of science and its hope for man. He left his work unfinished and open to improvement, as he knew science itself to be unfinished in the world of time. He warned man of the shadows in his own brain which kept him dancing like one enchanted within the little circle of narrow prejudice and fanatic ideology.

Though Bacon died childless, his intellectual children in a few scant generations were destined to be legion. He would be called by Izaak Walton a few years after his death "the great Secretary of all Nature." He planned as a man of great affairs, dreamed like a poet, and yet sought to rope those dreams to earth as though, in doing so, he might more easily sail earth itself into the full wind of that oncoming and creative age toward which his vision hastened.

We of today, cognizant of the earth's long past and our own bestial and half-human origins, grow fainthearted at the road we have still to travel. Bacon saw within our souls the possibility of a return to a Garden, a regaining of innocence through wisdom. We, by contrast, must seek a garden drawn from our own imperfect evolutionary substance. This is why our imaginings must be so carefully scrutinized lest we call the wrong world into being by the dark magic of ill-considered thought.

This is the reason also why Bacon placed such emphasis upon education. The teacher—this oft-scorned handler of young minds—is literally engaged in the most tremendous task imaginable. He is encouraging in the minds of the young Bacon's power to create a choice of worlds.

Perhaps in the end the Great Bringer of all things out of darkness may smile at our human prejudice, for, fallen angel or risen ape, the end the noblest intellects pursue is still the same: the opening of man's mind to his true dignity. Francis Bacon, the solitary discoverer of a continent in time, has placed his footprint where no man may follow by mechanical invention alone. In prayer Bacon himself had urged "that human things may not prejudice such as are divine, neither that from the unlocking of the gates of sense, and the kindling of a greater natural light, anything of incredulity or intellectual night may arise in our minds towards divine mysteries."

For us of this age the shadow of that night of which he spoke has fallen across half the world. Perhaps, however, he saw deeper, further, than ourselves, aware as he was of the "impediments and clouds in the mind of man." "I do not think ourselves," he ventured, without the twentieth century's boastful arrogance, "yet learned or wise enough to wish reasonably for man. I wait for harvest time, nor attempt to reap green corn."

3

The Orphic Theatre

IN THE COURSE OF THREE CENTURIES OF EXAMINA-
tion of Bacon's career, many of the ideas propounded about
him have been expressed, forgotten, or re-elaborated by
later writers. Kuno Fischer, in the mid-nineteenth century, dwelt upon his marvelous grasp of time. Similarly,
Richard ·Whately perceived that Bacon was a statesman
and strategist of science rather than an individual discoverer. Lord Acton, the well-known historian, somewhat
later recognized the value in Bacon's promotion of a division of labor in science by means of which discovery
need not always wait upon the chance appearance of
great genius. He saw the idea as equally applicable in the
field of historical research and did not project upon it
some of the misplaced animus of scientists jealous of their
role as geniuses. Most of these ideas of Bacon's have now
been realized in actual practice but have become so com-
monplace with the years that his farsighted role in their
encouragement has been forgotten, as well as the atmosphere of opposition in which they were produced.

Bacon was not a system builder save in his grasp of the
potential powers concealed in scientific research. In fact he

abjured dogmatic and full-formed philosophies, saying, in
the *Novum Organum*: "These I call Idols of the Theatre,
because in my judgment all the received systems are but so
many stage-plays, representing worlds of their own creation after an unreal and scenic fashion. . . . Many more
plays of the same kind may yet be composed and in like
artificial manner set forth."

Neither, he warned, was science itself free of tradition
and credulity. In this last view he was not mistaken, yet he
was tolerant to the point of suggesting, elsewhere, the ad-
visability of collecting and comparing all such philosophies for the grains of truth that might lie in them. He
had a clear grasp of the importance of the history of ideas,
and in this respect once more forecast a field of thought
which is only now receiving the attention it deserves.

As a consequence of this aversion to doctrine, light and
air blow through his windy and perpetually unfinished
building. As Whately remarked, he labored, unlike the
system builders, to make himself useless. He saw his efforts
not as primarily theoretical but as a necessary prelude to
experiment—a work which he, a solitary man whose hourglass of days was short, could only dimly adumbrate.

The way of science, as someone has observed, is the road
of the snail that shineth, though it is slow. Yet in these
centuries of weighing and reweighing, detraction and
praise, no one seems to have encountered Bacon the anthropologist. This is the more strange because foe and
friend alike have been quick to admit that he emerged
from an atmosphere where the astute judgment of men
and the flow of events were paramount. Bacon himself,
perhaps out of bitter self-questioning and disappointment,

referred to the world he inhabited as one of shadow rather than of light, yet it must be suspected that it was from the observation of men and custom, out of science practiced "like a lord chancellor," that Bacon drew his laboratory materials and literally produced his second world, or, rather, laid its foundations in the minds of men.

I have said that his building is huge beyond our imaginations, drafty and unfinished. Like all such monuments of genius, it is never truly of the past. Lights flicker mistily in its inner darkness; stones are still moved about by unseen hands. Somewhere within, there is a ghostly sound of hammering, of a work being done. The work is ours, the building is as we are shaping it, nor would Bacon have it otherwise. Since Bacon was a statesman and a pathfinder, no man quite escapes his presence in the haunted building of science, or the whispers of his approbation or unease. Occasionally the voice grows louder, as now, in our over-toppling part of the structure. To those who listen, the harsh Elizabethan line strikes once more like surf around the shores of his far-off New Atlantis, warning us of man's double nature and perhaps his fate, for Bacon did not hesitate to write: "Force maketh Nature more violent in the Returne."

It is often asserted, frequently by experimental scientists, that Bacon, at best, was a stimulating popularizer of new doctrines, but not a true discoverer. These remarks have always seemed to me an appalling underestimation of Bacon's role in the history of human thought. Ironically, he who defended the technicians has frequently been rejected by them. Such men fail to recognize that mere words can sometimes be more penetrating probes into the nature of the universe than any instrument wielded in a laboratory. They have ignored his vision of the very education of which they are the present-day beneficiaries. They have failed to understand, because he wrote before the rise of a professional jargon, the basic contributions in the realm of the social and biological sciences which have passed imperceptibly into our common body of thought. Bacon's "great machine," his system of induction applied to the natural world about us, has obscured the recognition of much that he observed about human culture and the sociological nature of discovery. The instruments of the mind, he observes justly, are as important as the instruments in the hand. He himself has suffered from the effects of his own observation that "great discoveries appear simple once they are made."

Unfortunately, of no human endeavor is this remark more true than in the domain of abstract thought. A shift of wordage, an alteration in the direction of historical emphasis, and the most profound ideas may emerge in a new garb with their parentage forgotten and ignored. Oft times dismissed or misunderstood by the physical scientist, Bacon's contributions in the social sciences were made, in a sense, too early. By the time these subjects had emerged as recognized disciplines, his far-reaching, anticipatory insights were submerged in a welter of new books and newer phrasing. A few examples in the anthropological domain may illustrate the way in which time erodes the memory of great achievements in thought. In saying this let us remember that the thinker does not work in a vacuum.

Bacon, too, in spite of an active and original cast of mind, took in and transmuted much that came to him from Greek and Roman sources. Unlike his associates, however, he saw the ancients in the dawning light of modernity, not as colossi whose achievements the living could not equal, but, rather, reduced to their true stature as men—men whose accomplishments the hoarded experience of succeeding generations might enable us to surpass.

"For at that time," Bacon wrote in *Valerius Terminus*, "the world was altogether home-bred, every nation looked little beyond their own confines or territories, and the world had no through lights then, as it hath had since by commerce and navigation, whereby there could neither be that contribution of wits one to help another, nor that variety of particulars for the correcting of customary conceits [ideas]." In Bacon's time such a view of the ancients was iconoclastic and heretical to an extent difficult to appreciate in the intellectual climate of today.

Perhaps one of the most obscurely hidden yet profound insights Bacon possessed revolves around his discovery of the *mundus alter,* the world unknown. "By the agency of man," contends Bacon, in what was then a bold and novel interposition of the human into the natural universe, "a new aspect of things, a new universe comes into view." Characterized in the sea-language of the great Elizabethan voyagers, this new universe is, in reality, what the modern anthropologist calls the world of culture, of human "art" with all its permutations and emergent quality.

Bacon is not content to subsist in the natural world as its exists, nor to drift aimlessly in history. The focal point of all his thinking is action, not system building. His new world to be brought out of time by human ingenuity interpenetrates and is interfused with the natural universe, yet remains a thing apart. Prospero's isle, the New Atlantis, can be brought out of nature only by the magic of the human mind directing and utilizing, rather than contesting against, nature. "The empire of man over things depends wholly on the arts and sciences. For we cannot command nature except by obeying her."

As he proceeds in the *Novum Organum* it is interesting to note that long before the rise of anthropology as a science, he had seen the social dynamic involved in cultural change and divorced it from simplistic explanations based on biology. He says: "Let only a man consider what a difference there is between the life of men in the most civilized province of Europe, and in the wildest and most barbarous districts of New India [the New World]; he will feel it be great enough to justify the saying that 'man is a god to man,' not only in regard of aid and benefit, but also by a comparison of condition. *And this difference comes not from the soil, not from climate, not from race, but from the arts*" (italics mine).

Three-and-a-half centuries ago Bacon thus curtly dismissed the racist doctrines that have hovered like an inescapable miasma ever since. In addition he had clearly recognized the role of culture and cultural conditioning in human affairs. "Custom," he notes, "is the principal magistrate of man's life." Its predominance "is every where visible. Men . . . do just as they have done before; as if they were dead images and engines moved only by the wheels of custom." It is well, he mused in the *Essays,* to stand upon the hill of Truth and, as one might observe from some point of vantage the waverings of an uncertain battle, to see, through the clean air of reason "the errors, and wanderings, and mists, and tempests in the vale below." "So always," he adds solemnly, "that this prospect be with pity."

In his symbolic treatment of the fable of Orpheus we may observe, as balancing the previous quotation, Bacon's statesmanlike recognition of the role played by culture in controlling the otherwise uninhibited behavior of man. Here the harp of Orpheus is Bacon's "magistrate."

"In Orpheus's theatre all beasts and birds assembled, and forgetting their several appetites, some of prey, some of game, some of quarrel, stood all sociably together listening to the airs and accords of the harp, the sound whereof no sooner ceased, or was drowned by some louder noise, but every beast returned to his own nature; wherein is aptly described the nature and condition of men; who are full of savage and unreclaimed desires of profit, of lust, of revenge, which as long as they give ear to precepts, to laws, to religion sweetly touched with eloquence, and persuasion of books, of sermons . . . so long is society and peace maintained; but if these instruments be silent, or sedition and tumult make them not audible, all things dissolve in anarchy and confusion."

The cultural tie—custom, in other words—subdues man to its strange music and holds back the expression of his wilder nature.

If we pursue Bacon's influence upon later scholars in anthropology, we come upon a supreme example which remains oddly undocumented. Sir James Frazer, the distinguished British student of primitive magic and religion, makes a great point of the fact that all over the world magical practice seems to resolve itself into two principles of thought: that things once in contact with each other continue to act at a distance, or, second, that like things produce like results—that is, that the imitation, with proper ceremonial, of a natural phenomenon by the magician will initiate, for example, a real rainstorm. Concealed behind these two widespread branches of magic, Frazer believed, was a single spurious "scientific" principle which was common to the primitive and untutored mind. That principle he proclaimed as a mistaken Law of Sympathy, the conception "that things act on each other at a distance through a secret sympathy." Magic, as viewed by Frazer, is thus a kind of falsely conceived science based upon a naïve projection of human desire upon the exterior universe.

In the *Sylva Sylvarum,* that curious miscellany to which Bacon devoted his last years, there is a considered discussion of magic in which Bacon refers to the "operations of sympathy, which the writers of natural magic have brought into an art or precept." Bacon then discusses what Frazer was to label contagious and imitative magic, both, of course, distinct products of the underlying principle of sympathetic magic. I do not wish to subtract one iota of credit from the stimulating ideas contained in Frazer's classic, *The Golden Bough,* but I believe that Frazer's formulation of the "laws" of magic is drawn, in part, from Francis Bacon. While Frazer makes one use of magical example from the *Sylva Sylvarum,* showing that he was aware of Bacon's pioneer precedence, nowhere, to my knowledge, does he indicate Bacon's prior discussion of the principle "that things act on each other at a distance through a secret sympathy." Yet of this same principle Bacon had written, "of things once contiguous or entire there should remain a transmission of virtue from the one to the other."

Naturally Bacon was involved in other matters and did not refine or draw out of this material all that Frazer in a later century would be able to do. One is a little saddened, however, that Frazer did not see fit to acknowledge the stimulus of that elder scholar whose doom it was to plant seeds which sprang from his work at points so distant in time and space that only the most meticulous would acknowledge their indebtedness. Bacon was right in his noble and elevating remark that books, like ships, pass through the seas of time and touch other minds in distant ages. It is troubling to our human wish for remembrance that the launcher of such frail vessels never sees upon what shore they land or whether his precious cargo is carried off by erudite smugglers. Broken by the winds and waves of other ages, what the author painstakingly accumulated may be picked up piecemeal as wreckage and distributed in other holds. His one compensation must lie in the hope that the cargo, if not the master, is saved. In this respect Bacon deserves re-examination. A few examples taken almost at random from his work may prove convincing to those who have tended to dismiss his creative powers.

In *The Advancement of Learning* Bacon, in one of his characteristic axiomatic statements, has remarked: "He that cannot contract the sight of his mind as well as disperse and dilate it, wanteth a great faculty." It does not take long to demonstrate that Bacon possessed this rare trait to a unique extent. In the second book of the *Novum Organum* he dwells upon the types of insensible change which tend to escape our observation. As he again says so aptly, one of the great drawbacks of untrained sense perceptions is that "they draw the lines of things with reference to man, and not with reference to the universe; and this is not to be corrected except by reason and a universal philosophy." He indicates that much of significance in nature takes place by almost imperceptible progression which the heedless fail to remark. In this observation he was laying the theoretical groundwork which underlies the whole domain of biology from evolution and genetics to embryology. In fact, he not alone dilates his mind to take in this important principle of natural history but, true to his own words, contracts the principle sufficiently to describe minute experiments which were later carried out by others, very possibly in some cases under his influence.

He says, for example, that the embryology of plants can be studied through the simple process of inspecting day by day the growth of sprouting seeds. Similarly he comments, "We should do the same with the hatching of eggs, in which case we shall find it easy to watch the process of vivification and organization, and see what parts are produced from the yolk, and what from the white of the egg, and other things." The Italian anatomist Marcello Malpighi published his *De formatione pulli in ovo* in 1673;

Hieronymus Fabricius's similar study, *De formatione ovi et pulli,* which appeared during Bacon's last years, does not predate the publication of the *Novum Organum.* In the *Sylva Sylvarum* one notes another aspect of Bacon's "contracted" mind at work. Probing speculatively into the nature of life, he notes that "the Nature of Things is commonly better perceived in Small, than in Great." The investigation of lower and more simple animals, he believes, is more apt to reveal the secrets of life than similar studies made in "Disclosing many Things in the Nature of Perfect Creatures, which in them lye more hidden." Again the modern biologist would be forced to agree.

Jean Rostand has paid tribute to Bacon's pioneer interest in teratology, "of all prodigies and monstrous births of Nature, of everything in short that is new, rare and unusual." Here, argues Bacon, nature in her errors reveals herself unbidden. The knowledge of life's mutations and deflections should aid us in comprehending the mysteries of life. The solution of those mysteries was many generations away, but, once more, that quick impatient brain had probed an instant into the domain of evolutionary change.

Archibald MacLeish, in a fine poem, "Epistle to Be Left in the Earth," narrates the thoughts of a last, desperate survivor on a dying earth. "I pray you," the unknown individual addresses hopefully a race to come:

Make in your mouths the words that were our names.

Laboring under the heavy burden of his own mortality, he cries anxiously:

I will tell you all we have learned
I will tell you everything.

The poem then runs off into the disjointed efforts of a
man scrawling, haphazard, a last few facts about this
planet:

It is colder now
        there are many stars
       we are drifting
North by the Great Bear
       the leaves are falling
the water is stone in the scooped rocks.

There is a poignant similarity between this verse and
the real-life creation of Bacon's *Sylva Sylvarum*. The book
was written in haste in his declining years after all his
hopes for new universities, co-workers in science, and aid
from enlightened rulers had been disappointed. But the
man, like the person in MacLeish's poem, still struggled to
collect the facts out of which the new continent should be
built. He is aging, hope is gone, the task looms gigantic.
He has no adequate conception of the size of the universe
he has attempted to engage. No one will come to his aid.
The book becomes an almost incoherent babble of facts
drawn both from personal observations and diverse
sources. Moss grows on the north side of trees. Strange
fungi spring up in the forests. Fruit put into bottles and
lowered into wells will keep long. He writes of the moon
and vinegar, of cuttle ink, and of the glowworm. These
are such facts as each one of us, divested of four centuries

of learning, might try to record for posterity if he were the last of a dying race.

In a sense Francis Bacon was such a man. He was dying seemingly without scientific issue; the great continuity of learning for which he pleaded had been received indifferently by the world. Yet hidden in the *Sylva Sylvarum,* regarded as of little importance today, is a quite remarkable statement.

The passage is striking because it sets the stage for as pure a demonstration of the value of the induction for which Bacon argued as he could possibly have hoped for. Yet because two hundred and fifty years were to elapse in the reasoning process, men have forgotten the connection. After some observations upon changes in plants he remarks:

"The transmutation of species is, in the vulgar philosophy, pronounced impossible, and certainly it is a thing of difficulty, and requireth deep search into nature; but seeing there appear some manifest instances of it, the opinion of impossibility is to be rejected, and the means thereof to be found out."

This tolerant and studious observation with its evolutionary overtones was made before the nature of fossils was properly understood and before the length of geological time had been appreciated.

"The path of science," Bacon had proclaimed, "is not such that only one man can tread it at a time. Especially in the collecting of data the work can first be distributed and then combined. Men will begin to understand their own strength only when instead of many of them doing the same things, one shall take charge of one thing and one of another."

For the next two hundred years men allied in international societies originally foreseen by Bacon would make innumerable observations upon the strata of the earth, upon fossils, and upon animal and plant distributions. Heaps upon heaps of facts collected and combined by numerous workers would eventually lead to Darwin's great generalization. In the end Darwin himself was to write, perhaps with a touch of guile, "I worked upon the true principles of Baconian induction." The individual empirical observations which led to the theory of evolution and the recognition of human antiquity had been wrenched piecemeal from the earth.

Bacon, moreover, was keenly cognizant of the value of the history of science and philosophy. He deplored its neglect and urged that all such history of "oppositions, decays, depressions [and] removes . . . and other events concerning learning will make men wise." Even further, in *The Advancement of Learning* he indicates a complete awareness of what today we would call the intellectual climate, or *Zeitgeist,* of an age. He recognizes, and knows within himself, that it is possible for able men to find themselves in an uncongenial age. He advises men "to consider how the constitution of their nature sorteth with the general state of the times." If their natures and that of the time are congenial they may allow themselves "more scope and liberty." Otherwise their lives must "be more close, retired, and reserved." There can be no doubt that Bacon, for all his role in public office, preferred consciously the latter course. His deep interest in history, his pioneer advocacy of studies in comparative government, his plea for personal biographies of men other than princes, all reveal the breadth of his horizon-circling mind.

Bacon also fully anticipated the folly of great thinkers in their tendency to extrapolate too broadly from the base of a single discovery. Thus, though he admired the discovery of the magnet by William Gilbert, physician to both Elizabeth and James, Bacon wrote: "he has himself become a magnet: that is he ascribed too many things to that force." One might observe that this well-known tendency is apparent in Louis Agassiz's final exaggeration of the extent of the Ice Age, so that he envisioned it as covering the Amazon Basin. Other equally pertinent examples could be cited. Even great thinkers, or if not they, then their followers, sometimes show a tendency to create anew a ring within which they dance. Bacon's own emphasis upon induction was to suffer from similar misuse by his followers of the early nineteenth century, who seized upon it as a device with which to castigate the evolutionists and geologists as "speculating" from insufficient accumulations of fact.

One other view which Bacon advocated was severely criticized by the nineteenth-century British essayist Thomas Babington Macaulay and still evokes comment today— namely, his so-called neglect of genius. "The course I propose for the discovery of sciences," Bacon argues, "is such as leaves but little to the acuteness and strength of wits, but places all wits and understandings nearly on a level."

At first glance such a remark is apt to offend the superior intellect. In actuality, however, just as Clemenceau is reputed to have remarked that war was too important to be left to generals, so Bacon is not content to leave the development of the sciences to the sporadic appearances of genius. Nor I think, practically speaking, would any educator today.

As I have already noted, a careful reading of Bacon reveals that what he is anxious to achieve is the triumph of the experimental method. This triumph demands the thorough institutionalization of science at many levels of activity. In one passage he encompasses in a brief fashion all those levels on which science operates today. "I take it," he writes prophetically, "that all those things are to be held possible and performable, which may be done by some persons, though not by everyone; and which may be done by many together, though not by one alone; and which may be done in the succession of ages, though not in one man's life; and lastly, which may be done by public designation and expense, though not by private means and endeavor." Only in this manner can the continuity of the scientific tradition be maintained and the small bricks which go into the building of great edifices be successfully gathered. "Even fourth-rate men," Darwin was later to observe, "I hold to be of very high importance at least in the case of science."

If Bacon meant anything at all, he meant that working with the clay that sticks to common shoes was the only way to ensure the emergence of order and beauty from the misery of common life as his age knew it. He eliminated, in effect, reliance upon the rare elusive genius as a safe road into the future. It partook of too much risk and chance to rely upon such men alone. One must, instead, place one's hopes for Utopia in the education of plain Tom Jones and Dick Thickhead.

Ironically, this was the message of a very great genius, an aristocrat who had lived all his life in the pomp of circumstance, but who, in the end, was willing to leave his name to later ages and his work to their just judgment. Bacon had an enormous trust in the capacities of the human mind, even though no one had defined better than he its idols and distortions. "There be nothing in the globe of matter," he wrote, "that has not its parallel in the globe of crystal or the understanding." John Locke, almost a century later, was far more timid than this. Perhaps Bacon reposed too much hope in the common man. Or perhaps it is we who lack hope—the age for which Bacon waited being still far off, or a dream. But is it not a very great wonder that a man who spent all his life in the arrogant class-conscious court of a brutal age strove for personal power as a means of transmitting to the future an art which would in a sense make him, a very great genius, and men like him less needed for human advancement?

I know of no similar event in all history. As an educator in a country which has placed its faith in the common man, I can only say that the serenity of Bacon's faith takes our breath away and gives him, at the same time, our hearts. For he, the Lord Chancellor, was willing to build his empire of hope from common clay—from men such as you and I. "It is not," he protests, "the pleasure of curiosity . . . nor the raising of the spirit, nor victory of wit, nor lucre of profession, nor ambition of honor or fame, nor inablement for business, that are the true ends of knowledge." Rather it is "a restitution and reinvesting of man to the sovereignty and power which he had in the first state of creation."

Those who dismiss Bacon as a scientist because he made no mechanical inventions have forgotten his own uncanny and preternatural answer, for he remarked that in the beginning there was only light. The grinding of machines and the sounds above us in the air we have taken for the scientific fruits he spoke of. After four hundred years the light, however, is only along the horizon—that beautiful, dry light of reason, which Bacon admired above all things, and which he spoke of as containing charity.

Ours cannot be the light he saw. Ours is still the vague and murky morning of humanity. He left his name, the name of all of us, to the charity of foreigners and the next ages. We presume if we think we are those addressed in his will. We are, instead, only a weary renewed version of the court he knew and the days he wore out in blackness. The inertial guidance system in the warhead has no new motive behind it; the Elizabethan intrigues that flung up men of power and destroyed them have a too familiar look; the religious massacres that shook Bacon's century have only a different name in ours.

There is something particularly touching about Bacon's growing concern "to make the mind of man, by help of art, a match for the nature of things." He knew, in this connection, that man the predator is also part of that nature man had to conquer in order to survive. Bacon had sat long in high places; he knew well men's lusts and rapacities. He knew them in the full violence of a barbaric age.

Although he has been accused of giving "good advice

for Satan's Kingdom," he understood from the beginning, and stated in no uncertain terms, that the technological arts "have an ambiguous or double use, and serve as well to promote as to prevent mischief and destruction, so that their virtue almost destroys or unwinds itself." "All natural bodies," he contended, with some dim, evolutionary foreboding, "have really two faces, a superior and inferior."

In one of those strange yet powerful sentences which project like reefs out of the sum total of his work, he gives us a Delphic prophecy: "Whatever vast and unusual swells may be raised in nature," he says, "as in the sea, the clouds, the earth or the like"—so that in this age our mind flies immediately to man—"yet nature," he continues, "catches, entangles and holds all such outrages and insurrections in her inextricable net, wove as it were of adamant."

John Locke, struck by the immensity of the great American forest, cried out, "In the beginning the whole world was like America." Had Bacon seen the possibility of the return of that forest even before it had departed? Or did he look beyond this age to a time when, by greater art than now we practice, we may have made our peace with the nature of things?

"It must ever be kept in mind," Bacon urged, "that experiments of Light are more to be sought after than experiments of Fruit." The man was obsessed by light—that pure light of the first morning of creation before the making of things had commenced, before there was a Garden and a serpent and a Fall, before there was strontium and the shadow of the mushroom cloud. He who will not attend to things like these can, in Bacon's own words, "neither win the kingdom of nature nor govern it."

Because Bacon saw and understood this light, it is well, I think, that he be not judged by us. Those who charge him with having, like a necromancer, called up from the deeps of time the direst features of the modern age, should ponder well his views upon the soul—"the world being in proportion," as he says, "inferior."

"By reason whereof," Bacon adds, "there is agreeable to the spirit of man, a more ample greatness, a more exact goodness, and a more absolute variety, than can be found in the nature of things."

The means to that goodness and those uses of life which Bacon sought for man can now be summarized. They are underlying precepts so firmly impressed upon our generation that we take them for granted, forgetting that, in the words of Kuno Fischer, "It was only Bacon's enthusiasm maintained through half a century, his dauntless tenacity . . . and his splendid powers of speech that gave to science wings."

First, Bacon wrote, "upon a given body to generate and superinduce a New Nature or Natures, is the work and aim of human power." He distinguished, in other words, between the lurking creativeness in nature and the natural world unsubjected to human influence. From "the beaten and ordinary paths of Nature" he would lead us to "another Nature which shall be convertible with the given Nature." Here, he indicates, lies a road to human power *"as in the present state of things human thought can scarcely comprehend or imagine"* (italics mine). Over and over in one form or another he draws a distinction between *natura libera* and *natura vexata*. The last, of course, is nature vexed, put to the question, examined by means of experiment.

Second, as is evident from the human intrusion into nature just exemplified, Bacon, more than any other man of his epoch, recognized not alone the concealed novelty residing in nature, but also the unexploited power of humanity, not just to *live* in nature, but to create a new nature through the right use of human reason—"that which," as he expressed it, "without art would not be done." In this intensely imaginative act he recognized the creativity also implicit in the properly disciplined mind. He saw how the universe could be ingested by the human brain and reflected back upon nature in new forms and guises. It is the great scope of this penetrating generalization, its recognition of the role of human culture both in nature and in history, which is so startling to discover at the dawn of the scientific era. This insight alone ranks Bacon far above the journeyman investigator, no matter how able or well-intentioned.

Last, and perhaps most important, Bacon had a powerful sense of time. He looked upon it in a new way, even if he could not foresee its ultimate extent. "It may be objected to me with truth," he pleaded against his own mortality, "that my words require an age, a whole age to prove them, and many ages to perfect them." It has been said of primitive man and could be said even of men not so primitive, such as our Elizabethan forebears, that they lacked the words and concepts to deal adequately with events remote in time or hidden behind the outer show of visible nature. It was Bacon's forceful defense of education

—education *future,* not static or directed toward the past, that eventually swung people's heads about in a new direction. He was literally forcing man to grow conscious of his own culture by projecting an ideal yet dynamic version of it into the future. Inventiveness, creativity, lay in the future, not the past.

Bacon's human world can be seen as lying somewhere between plain raw nature and Plato's world of disembodied ideal forms. It is the cultural world of man, subject, it is true, to confusion and contingency, but existing, so long as man exists, alongside of the physical world. This world is both disembodied and wonder-working. To produce its wonders it is only necessary for man in both reason and charity to turn his head toward the future rather than the past. Thus Bacon strives to make of man an actively anticipatory, rather than reminiscent or "present," creature. To anticipate, however, the human being must be made conscious of his own culture and the modes of its transmission. Education must assume a role unguessed in his time and imperfectly realized in ours. It must neither denigrate nor worship the past: it must learn from it.

"Make," Bacon wrote, "the time to come the disciple of the time past and not the servant." The words ring with such axiomatic brevity that without reflection the sharp axe blade of his thought glances aside from our dull heads. Yet in that single phrase is contained the spirit of Bacon, the educator who, though trained in the profession of law, admonished posterity, "Trust not to your laws for correcting the times but give all strength to good education." No man in the long history of thought strove harder to lay his hands upon the future for the sake of unborn generations.

The words connote the essence of all that Bacon dreamed. Strangely, they are incorporated, no doubt imperfectly, in the life of the nation whose first colonizers left England in the *Sarah Constant* the year *The Advancement of Learning* was published. New India, he called us then. America was mountains and savages and untamed horizons. Bacon's ideas would lodge there like flying seeds and find suitable soil in the wilderness. A "new" and hopeful continent would arise from minds at work in the forest—a "Newfound land," in Bacon's own words, "of inventions and sciences unknown."

4

Strangeness in the Proportion

"I MAY TRULY SAY," WROTE FRANCIS BACON in the time of his tragic fall in 1621, "my soul hath been a stranger in the course of my pilgrimage. I seem to have my conversation among the ancients more than among those with whom I live." I suppose, in essence, this is the story of every man who thinks, though there are centuries when such thought grows painfully intense, as in our own. Bacon's contemporary, Shakespeare, also speaks of it from the shadows when he says:

> Sir, in my heart there was a kinde of fighting,
> That would not let me sleepe.

In one of those strange, elusive stories upon which Walter de la Mare exerted all the powers of his marvelous poetic gift, a traveler musing over the quaint epitaphs in a country cemetery suddenly grows aware of the cold on a bleak hillside, of the onset of a winter evening, of the miles he has yet to travel, of the solitude he faces. He turns

91

to go and is suddenly confronted by a man who has appeared from no place the traveler can discover, and who has about him, though he is clothed in human garb and form, an unearthly air of difference. The stranger, who appears to be holding a forked twig like that which diviners use, asks of our traveler, the road. "Which," he queries, "is the way?"

The mundane, though sensitive, traveler indicates the high road to town. The stranger, with a look of revulsion upon his face, almost as though it flowed from some secret information transmitted by the forked twig he clutches, recoils in horror. The way—the human way—that the traveler indicates to him is obviously not his way. The stranger has wandered, perhaps like Bacon, out of some more celestial pathway.

When the traveler turns from giving directions, the stranger has gone, not necessarily supernaturally, for de la Mare is careful to move within the realm of the possible, but in a manner that leaves us suddenly tormented with the notion that our road, the road to town, the road of everyday life, has been rejected by a person of divinatory powers who sees in it some disaster not anticipated by ourselves. Suddenly in this magical and evocative winter landscape, the reader asks himself with an equal start of terror, "What *is* the way?" The road we have taken for granted is now filled with the shadowy menace and the anguished revulsion of that supernatural being who exists in all of us. A weird country tale—a ghost story if you will—has made us tremble before our human destiny.

Unlike the creatures who move within visible nature

and are indeed shaped by that nature, man resembles the changeling of medieval fairy tales. He has suffered an exchange in the safe cradle of nature, for his earlier instinctive self. He is now susceptible, in the words of theologians, to unnatural desires. Equally, in the view of the evolutionist, he is subject to indefinite departure, but his destination is written in no decipherable tongue.

For in man, by contrast with the animal, two streams of evolution have met and merged: the biological and the cultural. The two streams are not always mutually compatible. Sometimes they break tumultuously against each other so that, to a degree not experienced by any other creature, man is dragged hither and thither, at one moment by the blind instincts of the forest, at the next by the strange intuitions of a higher self whose rationale he doubts and does not understand. He is capable of murder without conscience. He has denied himself thrice over, and is as familiar as Judas with the thirty pieces of silver.

He has come part way into an intangible realm determined by his own dreams. Even the dreams he doubts because they are not fanged and clawed like the life he sees about him. He is tormented, and torments. He loves, and sees his love cruelly rejected by his fellows. Far more than the double evolutionary creatures seen floundering on makeshift flippers from one medium to another, man is marred, transitory, and imperfect.

Man's isolation is even more terrifying if he looks about at his fellow creatures and searches for signs of intelligence behind the universe. As Francis Bacon saw, "all things . . . are full of panic terrors; human things most of all; so

infinitely tossed and troubled as they are with superstition
[which is in truth nothing but a panic terror] especially
in seasons of hardship, anxiety, and adversity."

Unaided, science has little power over human destiny
save in a purely exterior and mechanical way. The beacon
light of truth, as Nathaniel Hawthorne somewhere remarks, is often surrounded by the flapping wings of un-
gainly night birds drawn as unerringly as moths toward
candlelight. Man's predicament is augmented by the fact
that he is alone in the universe. He is locked in a single
peculiar body; he can compare observations with no other
form of life.

He knows that every step he takes can lead him into
some unexplored region from which he may never return.
Each individual among us, haunted by memory, reveals
this sense of fear. We cling to old photographs and letters
because they comfort our intangible need for location in
time. For this need of our nature science offers cold comfort. To recognize this, however, is not to belittle the role
of science in our world. In his enthusiasm for a new magic,
modern man has gone far in assigning to science—his own
intellectual invention—a role of omnipotence not inherent in the invention itself. Bacon envisioned science as a
powerful and enlightened servant—but never the master
—of man.

One of the things which must ever be remembered
about Francis Bacon and the depth of his prophetic insight is that it remains, by the nature of his time, in a
sense paradoxical. Bacon was one of the first time-conscious moderns. He felt on his brow as did no other
man—even men more skilled in the devising of ex-

periment—the wind of the oncoming future, those far-off airs blowing, as he put it in the language of the voyagers, "from the new continent." Ironically, neither king, lawyer, nor scientist could tolerate Bacon's vision of the oncoming future.

That Bacon was a writer of great powers no one who has read his work would deny. He exercised, in fact, a profound stylistic influence both upon English writers who followed him and upon the scientists of the Royal Society. To say, for this reason, that he is of no scientific significance is to miss his importance as a statesman and philosopher of science as well as to deny to the scientist himself any greater role in discovery than the casual assemblage of facts. Harvey's attitude serves only to illustrate that great experimental scientists are not necessarily equally great philosophers, and that there may be realms denied to them. Similar able but particulate scientists, it could easily be pointed out, wrote disparagingly of Darwin in his time.

The great synthesizer who alters the outlook of a generation, who suddenly produces a kaleidoscopic change in our vision of the world, is apt to be the most envied, feared, and hated man among his contemporaries. Almost by instinct they feel in him the seed of a new order; they sense, even as they anathematize him, the passing away of the sane, substantial world they have long inhabited. Such a man is a kind of lens or gathering point through which past thought gathers, is reorganized, and radiates outward again into new forms.

"There are . . . minds," Ralph Waldo Emerson once remarked, "that deposit their dangerous unripe thoughts here and there to lie still for a time and be brooded in other minds, and the shell not to be broken until the next age, for them to begin, as new individuals, their career." Francis Bacon was such a man, and it is perhaps for this very reason that there has been visited upon him, by both moralist and scientist alike, so much misplaced vituperation and rejection.

He has been criticized, almost in the same breath, as being falsely termed a scientist and as being responsible for all the technological evils from which we suffer in the modern age. His vision was, to a degree, paradoxical. The reason lies in the fact that even the great visionary thinker never completely escapes his own age or the limitations it imposes upon him. Thus Bacon, the weather-tester who held up a finger to the winds of time, was trapped in an age still essentially almost static in its ideas of human duration and in the age and size of man's universe.

A man of the Renaissance, Bacon, for all his cynicism and knowledge of human frailty, still believed in man. He argues well and lucidly that to begin with doubt is, scientifically, to end in certainty, while to begin in certainty is to end in doubt. He failed to see that science, the doubter, might end in metaphysical doubt itself—doubt of the rationality of the universe, doubt as to the improvability of man. Today the "great machine" that Bacon so well visualized, rolls on, uncontrolled and infinitely devastating, shaking the lives of people in the remote jungles of Vietnam as it torments equally the hearts of civilized men.

It is evident from his *New Atlantis* that this attempt to picture for humanity the state it might attain under science and just rule retains a certain static quality. Bacon is sure about the scientific achievements of his ideal state, but, after all, his pictured paradise is an island without population problems, though medicine there is apparently a high art. Moreover, like most Utopias, it is hidden away from the corrupting influence of the world. It is an ideal and moving presentation of men going about their affairs under noble and uplifting circumstances. It is, as someone has remarked, "ourselves made perfect."

But as to how this perfection is to come into being, Bacon is obscure. It is obvious that the wise men of the New Atlantis must keep their people from the debasing examples of human behavior in the world outside. Bacon, in other words, has found it easier to picture the growth of what he has termed "experiments of fruit" than to establish the reality of a breed of men worthy to enjoy them. Even the New Atlantis has had to remain armed and hidden, like Elizabethan England, behind its sea fogs.

The *New Atlantis* cannot be read for solutions to the endless permutations and combinations of cultural change, the opened doorway through which Bacon and his followers have thrust us, and through which there is no return. To Bacon all possible forms of knowledge of the world might, he hoped, be accumulated in a few scant generations. With education the clouded mirror of the mind might be cleansed. "It is true," he admitted in an earlier work, *Valerius Terminus*, "that there is a limitation rather potential than actual which is when the effect is possible, but the time or place yieldeth not matter or basis whereupon man should work."

In this statement we see the modern side of Bacon's mind estimating the play of chance and time. We see it again a few pages later when, in dealing with the logical aspects of contingency, he writes, "our purpose is not to stir up men's hopes but to guide their travels." "Liberty," he continues, speaking in a scientific sense, "is when the direction is not restrained to some definite means, but comprehendeth all the means and ways possible." For want of a variety of scientific choice, he is attempting to say, you may be prevented from achieving a scientific good, some desirable direction down which humanity might travel. The bewildering multiplicity of such roads, the recalcitrance of even educated choice, are not solely to be blamed on Bacon's four-hundred-year gap in experience. In fact, it could readily be contended that science, as he intended to practice it, has not been practiced at all.

Although he has been hailed with some justice as the prophet of industrial science, it is often forgotten that he wished from the beginning to press forward on all scientific fronts at once, instead of pursuing the piecemeal emergence of the various disciplines in the fashion in which investigation was actually carried out. Three centuries have been consumed in establishing certain anthropological facts that he asserted from the beginning. He distinguished cultural and environmental influences completely from the racial factors with which they have been confused down to this day. He advocated the careful study and emulation of the heights of human achievement. Today scientific studies of "creativity" and the conditions governing the release of such energies in the human psyche are just beginning to be made. He believed and emphasized that it was within man's latent power to draw out of nature, as he puts it, "a second world."

It is here, however, that we come back upon that place
of numerous crossroads where man has lifted the lantern
of his intellect hopefully to many ambiguous if not treach-
erous sign posts. There is, we know now to our sorrow,
more than one world to be drawn out of nature. When
once drawn, like some irreplaceable card in a great game,
that world leads on to others. Bacon's "second world" becomes a multiplying forest of worlds in which man's ability to choose is subdued to frightened day-to-day decisions.

One thing, however, becomes ever more apparent: the
worlds drawn out of nature are human worlds, and their
imperfections stem essentially from human inability to
choose intelligently among those contingent and inter-
twined roads which Bacon hoped would enhance our
chances of making a proper and intelligent choice. Instead
of regarding man as a corresponding problem, as Bacon's
insight suggested, we chose, instead, to concentrate upon
that natural world which he truthfully held to be protean,
malleable, and capable of human guidance. Although
worlds can be drawn out of that maelstrom, they do not always serve the individual imprisoned within the substance
of things.

D'Arcy Thompson, the late renowned British naturalist,
saw, long after, in 1897, that with the coming of industrial
man, contingency itself is subjected to a kind of increasing
tempo of evolution. The simplicity of the rural village of
Shakespeare's day, or even the complex but stabilized and
harmonious life of a very ancient civilization, is destroyed
in the dissonance of excessive and rapid change. "Strike a
new note," said Thompson, "import a foreign element to

work and a new orbit, and the one accident gives birth to a myriad. Change, in short, breeds change, and chance—chance. We see indeed a sort of *evolution* of chance, an ever-increasing complexity of accident and possibilities. One wave started at the beginning of eternity breaks into component waves, and at once the theory of interference begins to operate." This evolution of chance is not contained within the human domain. Arising within the human orbit it is reflected back into the natural world where man's industrial wastes and destructive experiments increasingly disrupt and unbalance the world of living nature.

Bacon shared in some part with his age a belief in the biform nature of the worldly universe. "There is no nature," he says, "which can be regarded as simple; every one seeming to participate and be compounded of two." Man has something of the brute; the brute has something of the vegetable, the vegetable something of the inanimate body; and so, Bacon emphasizes, "all things are in truth biformed and made up of a higher species and a lower." Strange though it may seem, in this respect Bacon, though existing on the brief Elizabethan stage of a short-term universe, was perhaps better prepared for the protean writhings of external nature and the variability manifest in the interior world of thought than many a specialist in the physical and biological sciences who would follow him.

Patrick Cruttwell, in his study of Shakespeare, comments on how frequently war within the individual, a sense of divided personality, is widespread in the spirit of that age, as it also is in ours:

> Within my soul there doth conduce a fight
> Of this strange nature, that a thing inseparate,
> Divides more widely than the skie and earth.

How much more we would see, I sometimes think, if the world were lit solely by lightning flashes from the Elizabethan stage. What miraculous insights and perceptions might our senses be trained to receive amidst the alternate crash of thunder and the hurtling force that give a peculiar and momentary shine to an old tree on a wet night. Our world might be transformed interiorly from its staid arrangement of laws and uniformity of expression into one where the unexpected and blinding illumination constituted our faith in reality.

Nor is such a world as incredible as it seems. Physicists, it now appears, are convinced that a principle of uncertainty exists in the submicroscopic realm of particles and that out of this queer domain of accident and impact has emerged, by some kind of mathematical magic, the sustaining world of natural law by which we make our way to the bank, the theatre, to our homes, and finally to our graves. Perhaps, after all, a world so created has something still wild and unpredictable lurking behind its more sober manifestations. It is my contention that this is true, and that the rare freedom of the particle to do what most particles never do is duplicated in the solitary universe of the human mind.

The lightning flashes, the smashed circuits through which, on occasion, leaps the light of universes beyond our ken, exist only in rare individuals. But the flashes from such minds can fascinate and light up through the arts of

communication the intellects of those not necessarily en-
dowed with genius. In a conformist age science must, for
this reason, be wary of its own authority. The individual
must be re-created in the light of a revivified humanism
which sets the value of man the unique against that vast
and ominous shadow of man the composite, the predictable, which is the delight of the machine. The polity we
desire is that ever-creative polity which Robert Louis
Stevenson had in mind when he spoke of each person as
containing a group of incongruous and ofttimes conflicting citizenry. Bacon himself was seeking the road by
which the human mind might be opened to the full image
of the world, not reduced to the little compass of a state-
manipulated machine.

It is through the individual brain alone that there
passes the momentary illumination in which a whole
human countryside may be transmuted in an instant. "A
steep and unaccountable transition," Henry David Thoreau has described it, "from what is called a common sense
view of things, to an infinitely expanded and liberating
one, from seeing things as men describe them, to seeing
them as men cannot describe them." Man's mind, like the
expanding universe itself, is engaged in pouring over limitless horizons. At its heights of genius it betrays all the
miraculous unexpectedness which we try vainly to elimi-
nate from the universe. The great artist, whether he be
musician, painter, or poet, is known for this absolute unexpectedness. One does not see, one does not hear, until he
speaks to us out of that limitless creativity which is his
gift.

The flash of lightning in a single brain also flickers

along the horizon of our more ordinary heads. Without that single lightning stroke in a solitary mind, however, the rest of us would never have known the fairyland of *The Tempest*, the midnight world of Dostoevsky, or the blackbirds on the yellow harvest fields of Van Gogh. We would have seen blackbirds and endured the depravity of our own hearts, but it would not be the same landscape that the act of genius transformed. The world without Shakespeare's insights is a lesser world, our griefs shut more inarticulately in upon themselves. We grow mute at the thought—just as an element seems to disappear from sunlight without Van Gogh. Yet these creations we might call particle episodes in the human universe—acts without precedent, a kind of disobedience of normality, unprophesiable by science, unduplicable by other individuals on demand. They are part of that unpredictable newness which keeps the universe from being fully explored by man.

Since this elusive "personality" of the particle may play a role in biological change and diversity, there is a way in which the mysterious world of particles may influence events within the realm of the living. It is just here, within the human domain of infinite variability and the individual act, that the role of the artist lies. Here the creative may be contrasted to the purely scientific approach to nature, although we must bear in mind that a man may be both a scientist and artist—an individual whose esthetic and humanistic interests are as much a part of his greatness in the eyes of the world as the technical skills which have brought him renown.

Ordinarily, however, there is between the two realms a

basic division which has been widened in the modern world. Granted that the great scientific discoverer may experience the esthetic joy of the true artist, a substantial difference still remains. For science seeks essentially to naturalize man in the structure of predictable law and conformity, whereas the artist is interested in man the individual.

"This is your star," says science. "Accept the world we describe to you." But the escaping human mind cries out, in the words of G. K. Chesterton, "We have come to the wrong star. . . . That is what makes life at once so splendid and so strange. The true happiness is that we don't fit. We come from somewhere else. We have lost our way."

I once chanced to write a book in which I expressed some personal views and feelings upon birds, bones, spiders, and time, all subjects with which I had some degree of acquaintance. Scarcely had the work been published when I was sought out in my office by a serious young colleague. With utter and devastating confidence he had paid me a call in order to correct my deviations and lead me back to the proper road of scholarship. He pointed out to me the time I had wasted—time which could have been more properly expended upon my own field of scientific investigation. The young man's view of science was a narrow one, but it illustrates a conviction all too common today: namely, that the authority of science is absolute.

To those who have substituted authoritarian science for authoritarian religion, individual thought is worthless unless it is the symbol for a reality which can be seen, tasted, felt, or thought about by everyone else. Such men adhere to a dogma as rigidly as men of fanatical religiosity. They reject the world of the personal, the happy world of open, playful, or aspiring thought.

Here, indeed, we come upon a serious aspect of our discussion. For there is a widespread but totally erroneous impression that science is an unalterable and absolute system. It is supposed that other institutions change, but that science, after the discovery of the scientific method, remains adamant and inflexible in the purity of its basic outlook. This is an iron creed which is at least partly illusory. A very ill-defined thing known as the scientific method persists, but the motivations behind it have altered from century to century.

The science of the seventeenth century, as many historians have. pointed out, was essentially theoretical and other-worldly. Its observations revolved largely about a world regarded as under divine control and balance. As we come into the nineteenth century, cosmic and organic evolution begin to effect a change in religious outlook. The rise of technology gave hope for a Baconian utopia of the New Atlantis model. Problem solving became the rage of science. Today problem solving with mechanical models, even of living societies, continues to be popular. The emphasis, however, has shifted to power. From a theoretical desire to understand the universe, we have come to a point where it is felt we must understand it to survive. Governments expend billions upon particle research, cosmic-ray research, not because they have been imbued suddenly with a great hunger for truth, but for the very simple if barbarous reason that they know the power which lies in the particle. If the physicist learns the nature of the universe in his cyclotron, well and good, but the search is for power.

One period, for reasons of its own, may be interested in stability, another in change. One may prefer morphology, another function. There are styles in science just as in other institutions. The Christianity of today is not totally the Christianity of five centuries ago; neither is science impervious to change. We have lived to see the technological progress that was hailed in one age as the savior of man become the horror of the next. We have observed that the same able and energetic minds which built lights, steamships, and telephones turn with equal facility to the creation of what is euphemistically termed the "ultimate weapon."

It is in this reversal that the modern age comes off so badly. It does so because the forces which have been released have tended to produce an exaggerated conformity and, at the same time, an equally exaggerated assumption that science, a tool for manipulating the outside, the material universe, can be used to create happiness and ethical living. Science can be—and is—used by good men, but in its present sense it can scarcely be said to create them. Science in discovery represents the individual, but in the moment of triumph, science creates uniformity through which the mind of the individual once more flees away.

It is the part of the artist—the humanist—to defend that eternal flight, just as it is the part of science to seek to impose laws, regularities, and certainties. Man desires the certainties but he also transcends them. Thus, as in so many other aspects of life, man inhabits a realm half in and half out of nature, his mind reaching forever beyond the tool, the uniformity, the law, into some realm which is that of mind alone. The pen and the brush represent that eternal search, that conscious recognition of the individual as the unique creature beyond the statistic.

Modern science itself tacitly admits the individual, as in this statement from the English biologist P. B. Medawar: "We can be sure that, identical twins apart, each human being alive today differs genetically from any other human being; moreover, he is probably different from any other human being who has ever lived or is likely to live in thousands of years to come. The potential variation of human beings is enormously greater than their actual variation; to put it in another way, the ratio of possible men to actual men is overwhelmingly large."

So far does modern science spell out for us that genetic indeterminacy which parallels, in a sense, the indeterminacy of the subatomic particle. Yet all the vast apparatus of modern scientific communication seems fanatically bent upon reducing that indeterminacy as quickly as possible into the mold of rigid order. Programs which do not satisfy in terms of millions vanish from the air. Gone from most of America is the kind of entertainment still to be found in certain of the world's pioneer backlands where a whole village may gather around a little company of visitors. The local musician hurries to the scene, an artist draws pictures to amuse the children, stories are told with gestures across the barrier of tongues, and an enormous release of creative talent goes on into the small hours of the night.

The technology which, in our culture, has released

urban and even rural man from the quiet before his hearth log has debauched his taste. Man no longer dreams over a book in which a soft voice, a constant companion, observes, exhorts, or sighs with him through the pangs of youth and age. Today he is more likely to sit before a screen and dream the mass dream which comes from outside.

No one need object to the elucidation of scientific principles in clear, unornamental prose. What concerns us is the fact that there exists a new class of highly skilled barbarians—not representing the very great in science— who would confine men entirely to this diet. Once more there is revealed the curious and unappetizing puritanism which attaches itself all too readily to those who, without grace or humor, have found their salvation in "facts."

There has always been violence in the world. A hundred years ago the struggle for existence among living things was much written upon and it was popular for even such scholars as Darwin and Alfred Russel Wallace to dwell upon the fact that the vanquished died quickly and that the sum of good outweighed the pain. Along with the rising breed of scientific naturalists, however, there arose a different type of men. Stemming from the line of parson naturalists represented by Gilbert White, author of *The Natural History of Selborne*, these literary explorers of nature have left a powerful influence upon English thought. The grim portrait of a starving lark cracking an empty snail shell before Richard Jefferies' window on a bleak winter day is from a world entirely different from that of the scientist. Jefferies's observation is sharp, his facts accurate, yet there is, in his description, a sense of his own poignant hunger—the hunger of a dying man—for the beauty of an earth insensible to human needs. Here again we are in the presence of an artist whose vision is unique.

Even though they were not discoverers in the objective sense, one feels at times that the great nature essayists had more individual perception than their scientific contemporaries. Theirs was a different contribution. They opened the minds of men by the sheer power of their thought. The world of nature, once seen through the eye of genius, is never seen in quite the same manner afterward. A dimension has been added, something that lies beyond the careful analyses of professional biology. Something uncapturable by man passes over W. H. Hudson's vast landscapes. They may be touched with the silvery light from summer thistledown, or bleaker weathers, but always a strange nostalgia haunts his pages—the light of some lost star within his individual mind.

This is a different thing from that which some scientists desire, or that many in the scientific tradition appreciate, but without this rare and exquisite sensitivity to guide us the truth is we are half blind. We will lack pity and tolerance, not through intent, but from blindness. It is within the power of great art to shed on nature a light which can be had from no other source than the mind itself. It was from this doorway, perhaps, that de la Mare's celestial visitant had intruded. Nature, Emerson knew, is "the immense shadow of man." We have cast it in our image. To change nature, mystical though it sounds, we have to change ourselves. We have to draw out of nature that ideal second world which Bacon sought. The modern world is only slowly beginning to realize the profound implications of that idea.

Perhaps we can amplify to some degree certain of our observations concerning man as he is related to the natural world. In Western Europe, for example, there used to be a strange old fear, a fear of mountains, of precipices, of wild untrodden spaces which, to the superstitious heart, seemed to contain a hint of lurking violence or indifference to man. It is as though man has always felt in the presence of great stones and rarefied air something that dwarfed his confidence and set his thoughts to circling— an ice age, perhaps, still not outlived in the human mind.

There is a way through this barrier of the past that can be taken by science. It can analyze soil and stones. It can identify bones, listen to the radioactive tick of atoms in the lattices of matter. Science can spin the globe and follow the age-long marchings of man across the wastes of time and space.

Yet if we turn to the pages of the great nature essayists we may perceive once more the role which the gifted writer and thinker plays in the life of man. Science explores the natural world and thereby enhances our insight, but if we turn to the pages of *The Maine Woods*, regarded by critics as one of Thoreau's minor works, we come upon a mountain ascent quite unparalleled in the annals of literature.

The effect does not lie in the height of the mountain. It does not lie in the scientific or descriptive efforts made on the way up. Instead the cumulative effect is compounded of two things: a style so appropriate to the occasion that it evokes the shape of earth before man's hand had fallen upon it and, second, a terrible and original question posed on the mountain's summit. Somewhere along the road of that spiritual ascent—for it *was* a spiritual as well as a physical ascent—the pure observation gives way to awe, the obscure sense of the holy.

From the estimate of heights, of geological observation, Thoreau enters what he calls a "cloud factory" where mist was generated out of the pure air as fast as it flowed away. Stumbling onward over what he calls "the raw materials of a planet" he comments: "It was vast, titanic, and such as man never inhabits. Some part of the beholder, even some vital part, seems to escape through the loose grating of his ribs as he ascends. His reason is dispersed and shadowy, more thin and subtle, like the air. Vast, inhuman nature has got him at disadvantage, caught him alone, and pilfers him of some of his divine faculty." Thoreau felt himself in the presence of a force "not bound to be kind to man." "What is it," he whispers with awe, "to be admitted to a Museum, compared with being shown some star's surface, some hard matter in its home?"

At this moment there enters into his apprehension a new view of substance, the heavy material body he had dragged up the mountain the while something insubstantial seemed to float out of his ribs. Pausing in astonishment, he remarks: "I stand in awe of my body, this matter to which I am bound has become so strange to me. I fear not spirits, ghosts, of which I am one—*that* my body might—but I fear bodies, I tremble to meet them. What is this Titan that has possession of me? Talk of mysteries!—think of our life in nature—daily to be shown matter, to come in contact with it—rocks, trees, wind on our cheeks! the solid earth, the actual world." Over and over he muses, his hands on the huge stones, "*Who* are we? Where are we?"

The essayist has been struck by an enormous paradox. In that cloud factory of the brain where ideas form as tenuously as mist streaming from mountain rocks, he has glimpsed the truth that mind is locked in matter like the spirit Ariel in a cloven pine. Like Ariel, men struggle to escape the drag of the matter they inhabit, yet it is spirit that they fear. "A Titan grasps us," argues Thoreau, confronting the rocks of the great mountain, a mass solid enough not to be dragged about by the forces of life. "Think of our life in nature," he reiterates. "Who are we?"

From the streaming cloud-wrack of a mountain summit, the voice floats out to us before the fog closes in once more. In that arena of rock and wind we have moved for a moment in a titanic world and hurled at stone titanic questions. We have done so because a slight, gray-eyed man walked up a small mountain which, by some indefinable magic, he transformed into a platform for something, as he put it, "not kind to man."

I do not know in the whole of literature a more penetrating expression of the spirit's horror of the substance it lies trapped within. It is the cry of an individual genius who has passed beyond science into a high domain of cloud. Let it not be forgotten, however, that Thoreau revered and loved true science, and that science and the human spirit together may find a way across that vast mountain whose shadow still looms menacingly above us.

"If you would learn the secrets of nature," Thoreau in-sisted, "you must practice more humanity than others." It is the voice of a man who loved both knowledge and the humane tradition. His faith has been ill kept within our time.

Mystical truths, however, have a way of knowing nei-ther time nor total death. Many years ago, as an impres-sionable youth, I found myself lost at evening in a rural and obscure corner of the United States. I was there be-cause of certain curious and rare insects that the place af-forded—beetles with armored excrescences, stick insects which changed their coloration like autumn grass. It was a country which, for equally odd and inbred reasons, was the domain of people of similar exuberance of character, as though nature, either physically or mentally, had pre-pared them for odd niches in a misfit world.

As I passed down a sandy backwoods track where I hoped to obtain directions from a solitary house in the dis-tance, I was overtaken by one of the frequent storms that blow up in that region. The sky turned dark and a splatter of rain struck the ruts of the road. Standing uncertainly at the roadside I heard a sudden rumble over a low plank bridge beyond me. A man high on a great load of hay was bearing down on me through the lowering dark. I could hear through the storm his harsh cries to the horses. I stepped forward to hail him and ask directions. Perhaps he would give me a ride.

There happened then, in a single instant, one of those flame-lit revelations which destroy the natural world for-ever and replace it with some searing inner vision which accompanies us to the end of our lives. The horses, in the sound and fury of the elements, appeared, even with the loaded rick, to be approaching at a gallop. The dark figure of the farmer with the reins swayed high above them in some limbo of lightning and storm. At that moment I lifted my hand and stepped forward. The horses seemed to pause—even the rain.

Then, in a bolt of light that lit the man on the hayrick, the waste of sodden countryside, and what must have been my own horror-filled countenance, the rain plunged down once more. In that brief, momentary glimpse within the heart of the lightning, haloed, in fact, by its wet shine, I had seen a human face of so incredible a nature as still to amaze and mystify me as to its origin. It was—by some fantastic biological exaggeration—two faces welded vertically together along the midline, like the riveted iron toys of my childhood. One side was lumpish with swollen and malign excrescences; the other shone in the blue light, pale, ethereal, and remote—a face marked by suffering, yet serene and alien to that visage with which it shared this dreadful mortal frame.

As I instinctively shrank back, the great wagon leaped and rumbled on its way to vanish at what spot I knew not. As for me, I offer no explanation for my conduct. Perhaps my eyes deceived me in that flickering and grotesque darkness. Perhaps my mind had spent too long a day on the weird excesses of growth in horned beetles. Nevertheless I am sure that the figure on the hayrick had raised a shielding hand to his own face.

One does not, in youth, arrive at the total meaning of such incidents or the deep symbolism involved in them. Only if the event has been frightening enough, a revelation out of the heavens themselves, does it come to dominate the meaning of our lives. But that I saw the double face of mankind in that instant of vision I can no longer doubt. I saw man—all of us—galloping through a torrential landscape, diseased and fungoid, with that pale half-visage of nobility and despair dwarfed but serene upon a twofold countenance. I saw the great horses with their swaying load plunge down the storm-filled track. I saw, and touched a hand to my own face.

Recently it has been said by a great scientific historian that the day of the literary naturalist is done, that the precision of the laboratory is more and more encroaching upon that individual domain. I am convinced that this is a mistaken judgment. We forget—as Bacon did not forget—that there is a natural history of souls, nay, even of man himself, which can be learned only from the symbolism inherent in the world about him.

It is the natural history that led Hudson to glimpse eternity in some old men's faces at Land's End, that led Thoreau to see human civilizations as toadstools sprung up in the night by solitary roads, or that provoked Melville to experience in the sight of a sperm whale some colossal alien existence without which man himself would be incomplete.

"There is no Excellent Beauty that hath not some strangeness in the Proportion," wrote Bacon in his days of insight. Anyone who has picked up shells on a strange beach can confirm his observation. But man, modern man, who has not contemplated his otherness, the multiplicity of other possible men who dwell or might have dwelt in him, has not realized the full terror and responsibility of existence.

It is through our minds alone that man passes like that swaying furious rider on the hayrick, farther and more desperately into the night. He is galloping—this twofold creature whom Francis Bacon glimpsed—across the storm-filled heath of time, from the dark world of the natural toward some dawn he seeks beyond the horizon.

Across that midnight landscape he rides with his toppling burden of despair and hope, bearing with him the beast's face and the dream, but unable to cast off either or to believe in either. For he is man, the changeling, in whom the sense of goodness has not perished, nor an eye for some supernatural guidepost in the night.

Bibliography
Index

Bibliography

Anderson, Fulton H. *Francis Bacon: His Career and Thought*. Los Angeles: University of Southern California Press, 1962.

————. *The Philosophy of Francis Bacon*. Chicago: University of Chicago Press, 1948.

Bacon, Francis. *The Advancement of Learning*. Edited by William Aldis Wright. 5th ed. London: Oxford University Press, 1930. Introduction by the editor.

————. *Bacon's Essays and Annotations*. Edited by Richard Whately. 4th ed. London, 1858. Introduction by the editor.

————. *The Philosophical Works of Francis Bacon*. Edited by John M. Robertson. New York: Dutton, 1905.

Barnes, Harry Elmer. "The Philosophy of Francis Bacon," *Scientific Monthly* 18(1924), 475–495.

Baum, Bernard. "The Baconian Mind in Early Nineteenth Century America." Doctoral dissertation, University of Michigan, 1942.

Baumer, Franklin L. *Religion and the Rise of Scepticism*. New York: Harcourt, Brace and Co., 1960.

119

Bevan, Bryan. *The Real Francis Bacon.* London: Centaur Press, 1960.

Bierman, Judah. "New Atlantis Revisited," *Studies in the Literary Imagination* (Atlanta, Ga.), vol. 4, no. 1 (1971), 121–141.

Broad, C. D. *The Philosophy of Francis Bacon.* Cambridge, England: Cambridge University Press, 1926.

Crowther, J. G. *Francis Bacon: The First Statesman of Science.* London: Cresset Press, 1960.

Dixon, William Hepworth. *Personal History of Lord Bacon.* Boston: Ticknor and Fields, 1861.

Farrington, Benjamin. *Francis Bacon: Philosopher of Industrial Science.* New York: Schuman, 1949.

———. "Francis Bacon After His Fall," *Studies in the Literary Imagination* (Atlanta, Ga.), vol. 4, no. 1 (1971), 143–158.

Fischer, Kuno. *Bacon: His Philosophy and Time.* London: 1857.

Forsyth, A. R. "Address to the Mathematical and Physical Section of the British Association for the Advancement of Science," *Science,* n.s., 22(1905), 234–246.

Frazer, Sir James G. *The Golden Bough.* 1 vol. ed. New York: Macmillan, 1942.

Frost, Walter. *Bacon und die Naturphilosophie.* Munich: Ernst Reinhardt, 1927.

Fulton, J. F. "The Rise of the Experimental Method: Bacon and the Royal Society of London," *Yale Journal of Biology and Medicine,* March 1931.

Harris, Victor. *All Coherence Gone.* Chicago: University of Chicago Press, 1949.

Herschel, Sir John. *A Preliminary Discourse on the Study of Natural Philosophy.* Philadelphia: Carey, Lea and Blanchard, 1835.

Hill, Christopher. *Intellectual Origins of the English Revolu-
tion*. London: Oxford University Press, 1965.
————. *Puritanism and Revolution: Studies in Interpretation
of the English Revolution of the 17th Century*. London:
Secker and Warburg, 1958.
Hoopes, Robert T. *Right Reason in the English Renaissance*.
Cambridge, Mass.: Harvard University Press, 1962.
Lovejoy, Arthur O. *The Great Chain of Being: A Study of the
History of an Idea*. Cambridge, Mass.: Harvard University
Press, 1942.
Luxembourg, Lilo K. *Francis Bacon and Denis Diderot, Phi-
losophers of Science*. New York: Humanities Press, 1967.
Macaulay, Thomas Babington. "Lord Bacon," in *Critical,
Historical, and Miscellaneous Essays*, vol. III, 6 vols. New
York: Sheldon & Co., 1862.
Mathew, David. *Sir Tobie Mathew*. London: Max Parrish,
1950.
Medawar, P. B. *The Future of Man*. New York: Mentor Books,
1961.
Merton, Robert K. "Singletons and Multiples in Scientific
Discovery: A Chapter in the Sociology of Science," *Pro-
ceedings of the American Philosophical Society* (Phila-
delphia) 105(1961), 470–486.
Nichol, John. *Francis Bacon: His Life and Philosophy*. 2 vols.
London: Blackwood, 1901.
Penrose, Stephen B. L., Jr. "The Reputation and Influence of
Francis Bacon in the 17th Century." Doctoral dissertation,
Columbia University, 1934.
Prior, Moody E. "Bacon's Man of Science," *Journal of the
History of Ideas* 15(1959), 348–370.
Purver, Margery. *The Royal Society: Concept and Creation*.
London: Routledge and Kegan Paul, 1967.

Rostand, Jean. *Error and Deception in Science.* London: Hutchinson, 1960.

Schofield, Robert E. *Mechanism and Materialism: British Natural Philosophy in an Age of Reason.* New Haven, Conn.: Yale University Press, 1969.

Smith, David Nichol. *Characters from the Histories and Memoirs of the Seventeenth Century.* London: Oxford University Press, 1918.

Smith, Lacey Baldwin. "English Treason Trials and Confessions in the Sixteenth Century," *Journal of the History of Ideas* 15(1954), 471–498.

Spedding, James. *Evenings with a Reviewer, or Macaulay and Bacon.* 2 vols. London: 1881.

Storck, John. "Francis Bacon and Contemporary Philosophical Difficulties," *Journal of Philosophy* 28(1931), 169–186.

Thompson, Ruth D'Arcy. *D'Arcy Wentworth Thompson: The Scholar Naturalist 1860–1948.* London: Oxford University Press, 1958.

Tuveson, Ernest Lee. *Millennium and Utopia.* Berkeley, Calif.: University of California Press, 1949.

Westfall, Richard S. *Science and Religion in Seventeenth Century England.* New Haven, Conn.: Yale University Press, 1958.

Whewell, William. *On the Philosophy of Discovery.* London: John W. Parker and Son, 1860.

Williams, Charles. *Bacon.* New York, Harper & Brothers, 1933.

Wright, Jonathan. "Bacon and the Novum Organum," *Scientific Monthly* 26(1928), 34–40.

Index

123

About the Author

Loren Eiseley is Benjamin Franklin Professor of Anthropology and the History of Science at the University of Pennsylvania in Philadelphia, as well as being Curator of Early Man in the University Museum. Author of several award-winning books, Dr. Eiseley is widely known both as a naturalist and as a humanist. His work is represented in many anthologies of English prose, and he has the distinction of being an elected member of the National Institute of Arts and Letters. Dr. Eiseley has lectured at leading institutions of learning throughout the United States and has been the recipient of many honorary degrees.